The Unknown Component Problem

Tiziano Villa • Nina Yevtushenko
Robert K. Brayton • Alan Mishchenko
Alexandre Petrenko
Alberto Sangiovanni-Vincentelli

The Unknown Component Problem

Theory and Applications

Springer

Tiziano Villa
Dipartimento D'Informatica
Università di Verona
Strada le Grazie, 15
37134 Verona
Italy
tiziano.villa@univr.it

Robert K. Brayton
Department of Electrical Engineering
and Computer Science
University of California
Berkeley California
USA
brayton@eecs.berkeley.edu

Alexandre Petrenko
Computer Research Institute
of Montreal (CRIM)
Ogilvy 405, Suite 101
H3N 1M3 Montreal (Quebec)
Canada
petrenko@crim.ca

Nina Yevtushenko
Department of EECS
Tomsk State University
Tomsk
Russia
ninayevtushenko@yahoo.com

Alan Mishchenko
Department of Electrical Engineering
and Computer Science (EECS)
University of California, Berkeley
Berkeley California
USA
alanmi@eecs.berkeley.edu

Alberto Sangiovanni-Vincentelli
Department of Electrical Engineering
and Computer Science (EECS)
University of California, Berkeley
Berkeley California
USA
alberto@eecs.berkeley.edu

ISBN 978-0-387-34532-1 e-ISBN 978-0-387-68759-9
DOI 10.1007/978-0-387-68759-9
Springer New York Dordrecht Heidelberg London

Library of Congress Control Number: 2011938478

Printed on acid-free paper

Springer is part of Springer Science+Business Media (www.springer.com)

Mariae, uxori suavissimae, cum gratitudine: ad multos annos !

Tiziano

To my parents who gave me the courage and strength to go through life; to my family for their love and support

Nina

To my family, Ruth my wife of 50 years, children Jane Burchard, James and Michael Brayton, in-laws Dan Burchard and Debbie Marr, and grandchildren Thomas and Lia Burchard, and Matthew, Mason, Mariella and Marshall Brayton

Bob

To my family

Alan

To my parents

Alexandre

To my beloved Marica

Alberto

Preface

An important step in the design of complex systems is the decomposition of a system into a number of separate components which interact in some well-defined way. A typical question is how to design a component that when combined with a known part of a system, called the context, satisfies a given specification. This question arises in several applications ranging from logic synthesis to the design of discrete controllers. To formally solve such problems, the following questions need to be addressed:

- How to model the system, its components and the specification?
- How is the interaction between the components defined?
- When does a system behavior satisfy its specification?

For the first issue, different types of mathematical models can be used to specify the components of a system: finite automata (FA) and finite state machines (FSMs), ω-automata, Petri Nets are some of the most common formal models. Given an answer to the first question, matching answers must be provided to the other two.

For instance, if FSMs are used to model the system, operators to compose FSMs must be introduced together with the notion of an FSM conforming to another FSM. For FSM conformance, appropriate choices are language containment or simulation of one FSM by the other. For FSM composition, various forms have been described in the literature. For example, one can define an abstract equation over FSMs of the type $M_A \odot M_X \approx M_C$, where M_A models the context, M_C models the specification, M_X is unknown, $\odot$ stands for a composition operator (e.g., $\bullet$, synchronous composition or $\diamond$, parallel composition), and $\approx$ stands for a conforming relation (e.g., $\preceq$ for reduction relation, or $\cong$ for equivalence relation).

For any given formal model, appropriate equations can be set up and their solutions investigated. More complex equations or systems of equations can be formulated depending on the topology of the system's components.

In this book, we cast the problem of computing the unknown component in the common frame of solving equations over languages and automata. This allows to unify the treatment of a panoply of variants of this problem, as formulated by different research communities in different applicative domains. In particular, we treat in-depth equations over languages, regular languages, finite automata, finite state machines, and ω-automata. Then we apply the machinery of equations over finite state machines to sequential synthesis and resynthesis, and study a host of specific topologies, exact and heuristic techniques, and optimization scenarios. Finally, we enlarge the scope to domains such as testing, supervisory control, game theory, and specialized forms of synthesis for co-Büchi specifications. In the case of supervisory control, we characterize all controllers that satisfy partial controllability in the classical sense, and then we extend them to the relaxed notion of weak controllers.

The book is a blend of theory, especially in the chapters of Part I, with the description of a software package implementing most of the theory in Part II, together with applications to sequential synthesis in Part III and to sundry domains in Part IV.

This book grew out of an intense collaboration of the authors, in North America, Europe, and Russia. In particular, it capitalizes on the research activities carried forth in the CAD group at UC Berkeley, devoted to efficient optimization of digital designs, and on the theoretical research in Tomsk about automata theory. The convergence of these lines of investigation fostered a better understanding of the subject matter.

We thank all the Ph.D. students and researchers who contributed to this research in Berkeley, Tomsk and elsewhere, collaborating with the authors.

We are grateful to NATO for the funding of travels and meetings where the authors could converge from their distant locations, and where the book took shape. The NATO grants were:

1. "Logic synthesis and analysis through automata and language equation solving", NATO Science Program, No. PST.CLG.979698, Collaborative Linkage Grant. No. 971217, project duration: 2003–2005.
2. "Discrete Event System Optimization through Automata/FSM Equation Solving" NATO Collaborative Linkage Grant CBP.NR.CLG 982314, project duration: 2006–2009.

Among the participants to those meetings, we thank Roland Jiang, National Taiwan University, Taipei, and Anatoly Chebotarev, Ukrainian Academy of Sciences, Kiev, for interesting discussions on the themes covered in the book. A couple of meetings were held at PARADES, in the historical Palazzo Bonadies, Rome, and we thank Alberto for making that charming venue available to us. Tiziano thanks Bob for inviting him to visit UC Berkeley throughout many summers, to work on the book and enjoy the research ambience of his alma mater.

Thanks to Giovanni Castagnetti and Matteo Piccolo, research assistants at the University of Verona, for proofreading parts of the manuscript, and checking many of the examples with BALM. Giovanni and Matteo extended BALM (now BALM-II) to solve automatically parallel equations.

Verona	Tiziano Villa
Tomsk	Nina Yevtushenko
Berkeley	Robert K. Brayton
Berkeley	Alan Mishchenko
Montreal	Alexandre Petrenko
Berkeley	Alberto Sangiovanni-Vincentelli

Contents

Chapter 1
Synopsis

This book contains 18 chapters organized in four parts along with the synopsis and conclusion.

Each chapter is ended by a section of problems, that vary from testing the reader's grasp of the theory, to laboratory exercises with the software BALM, up to more open-ended problems at the level of a class project. This makes the book suitable also for advanced undergraduate and graduate teaching, besides being a reference for researchers.

Part I, Theory of Equations over Languages and Automata, presents the basics of equations over languages and automata. It is partitioned into four chapters.

In Chap. 2 we define and study abstract equations over languages, where a composition operator involved in the equations is either synchronous composition, $\bullet$, or parallel composition, $\diamond$, and the languages are compared using the language containment relation. The composition operators reflect two different types of interactions between systems found, e.g., in hardware and communicating systems. We consider thus two types of equations $A \bullet X \subseteq C$ and $A \diamond X \subseteq C$, where X represents an unknown language, while A and C represent known languages. The two types of equations are called, respectively, synchronous and parallel equations. A key contribution is the computation of the most general solutions of the language equations, found, respectively, as $S = \overline{A \bullet \overline{C}}$, and $S = \overline{A \diamond \overline{C}}$. These formulas turn out to subsume a panoply of specialized solutions derived in the past for specific composition operators and topologies. We then specialize language equations to regular languages, which are counterparts of finite automata. The idea is to operate algorithmically on regular languages via their automata to solve effectively the associated language equations. We conclude the chapter by applying the developed theory to a so-called converter synthesis problem, which occurs in the presence of different protocols.

In Chap. 3 we specialize language equations to FSM languages, counterparts of FSMs. We define synchronous and parallel compositions of FSMs and elaborate procedures for solving equations over FSMs. We also study subsets of solutions characterized by additional properties of practical interest, such as solutions that

T. Villa et al., *The Unknown Component Problem: Theory and Applications*,
DOI 10.1007/978-0-387-68759-9_1, © Springer Science+Business Media, LLC 2012

satisfy the Moore property and solutions to bounded parallel equations over FSMs, when external outputs should appear after a predefined number of internal actions.

In Chap. 4 we extend the results obtained in the finite regular case to equations over ω-automata. In this section, we explain how to extend parallel equations to ω-languages described by Büchi automata. We define expansion and restriction of ω-languages along with the corresponding ω-expansion and ω-restriction of ω-automata; we lift operations on regular languages to ω-regular languages and define ω-parallel composition of ω-automata. The largest solution of a parallel equation over Büchi automata is then formulated similarly to the finitary regular case.

Concluding Part 1, we provide in Chap. 5 an overview of various previous contributions, concerning specific aspects of the theory of synchronous and parallel equations over languages and automata. Synchronous equations naturally model numerous problems related to sequential circuits, such as hierarchical optimization, don't care conditions, flexibility and model matching. Equally, parallel equations offer a uniform way of modeling various domain problems, such as protocol converters and supervisory control. The overview indicates that the theory of synchronous and parallel equations models naturally a larger spectrum of language equations and their specialized solutions in different settings, since it is built upon the primitive notions of language and language composition. On the other hand, equation solving can be seen also as a special case of realizability/synthesis. Recall that given a logic formula, a classic problem of realizability is to determine whether there exists a sequential function to implement it, and if yes, find an implementation, which is a problem of synthesis. Problems of realizability/synthesis can be framed as solving games.

In Part II, Algorithms for Solving FSM Equations – BALM, we focus on the efficient implementation of operations and an environment suited for manipulating FSMs and solving equations over them. The developed software system, BALM, the Berkeley Automata and Language Manipulation system, is based on the existing multi-valued logic synthesis environment, MVSIS. This part is partitioned into four chapters.

In Chap. 6 we present several operations on finite automata implemented within BALM, which are used for solving equations. These operations include completion, determinization, complementation, product computation and others using a mixed representation of finite automata as state transition graphs, where transition predicates, if any, are represented by BDDs.

In Chap. 7 we outline the implementation of operations in BALM needed to solve equations over FSMs represented as sequential circuits. We discuss how a partitioned representation of sequential circuits can be used to perform the basic operations of equation solving: completing, complementing, determinizing, hiding, and computing the product of finite automata. Such a representation uses a set of next state functions, which update the state of each latch (flip-flop or state-holding element) in the circuit. We conclude the chapter by presenting experimental results which indicate that using a partitioned representation for FSM manipulations is more efficient compared to a monolithic representation.

Chapter 8 is devoted to the BALM package that includes the most typical automata operations, such as determinization and state minimization, as well as visualization capabilities, which rely on the graph visualization software Graphviz. We present and illustrate the two file formats supported in BALM, one for describing FSMs as multi-level netlists of logic, and the other for describing automata. We conclude the chapter by presenting a summary of the commands available in BALM, which can be used to program complex manipulations with finite automata and FSMs.

Chapter 9 provides a number of examples illustrating the use of BALM commands.

In Part III, Application to Sequential Synthesis, we consider the problem of sequential synthesis where a finite state machine is embedded in a larger FSM environment. We want to find the set of all possible FSMs that can replace the current FSM without changing the external behavior of the whole system. This set represents the complete sequential flexibility of the FSM (with respect to its environment); it is a generalization of the complete flexibility used in combinational synthesis. We also provide a number of examples of using BALM scripts in sequential synthesis. This part includes five chapters, the first four chapters address the problem of computing the flexibility in sequential networks; the last chapter discusses how to exploit the computed sequential flexibility by extracting a valid replacement with a chosen cost function.

Chapter 10 shows how BALM scripts implement several methods for computing a restricted form of sequential flexibility, sequential don't cares.

In Chap. 11, we explain how complete sequential flexibility for FSMs represented by netlists can be computed by windowing within BALM.

Chapter 12 addresses the problem of resynthesizing the component FSMs in a network of FSMs. We discuss the possibility of resynthesizing them one by one until the whole network cannot be simplified further. In order to avoid construction of the FSM that specifies the behavior of the whole network, we may consider just pairs of connected FSMs and find for a component of each pair a replacement by solving the corresponding equation at the price of missing some flexibility. Such a windowing approach is not feasible for all the topologies of FSM networks; for topologies, where some component's outputs are inputs of several components, we need to consider a system of FSM equations each of which involves a pair of such components. An FSM is a solution of a system of FSM equations if it is a solution of each equation of the system. Therefore the largest solution of a system of FSM equations can be obtained as the intersection of the largest solutions over all equations of the system.

Chapter 13 discusses solving language equations using simulation relations. The need for simulation relations arises in the so-called model matching problem. It asks for the design of a controller so that the composition of a plant with the controller matches a given model, as discussed in Chap. 5. The use of the simulation relation instead of the language containment relation allows us to obtain restricted solutions to FSM equations; we illustrate the possibilities of this approach with an example, where the BALM's command `latch_split` is used to divide the latches of a given FSM network into two parts, fixed and unknown.

Chapter 14 discusses the problem of selecting a "minimum" reduction or submachine of the largest solution and some criteria for selecting it given the flexibility computed by techniques presented in the previous chapters. We consider two criteria: the minimum number of states and the minimum input support. We recapture the basic notions of FSM state minimization, a classical problem studied for deterministic FSMs first and then for observable FSMs, and present a heuristic procedure for minimizing an FSM implemented in BALM. The criterion of minimum input support can be used when, in synthesizing the components of a synchronous composition of FSMs, it is desired to minimize not the number of states, but the number of communication lines of a given component. We present a procedure for deriving a largest solution that uses a minimal number of input alphabets (representing communication lines).

Part IV, More Applications of the Unknown Component Problem, comprises four chapters.

Chapter 15 focuses on supervisory control and explains the applicability of equation solving techniques to this theory. Methods for solving equations over regular languages have an exponential complexity in the worst-case, because of the complementation operator that requires determinization, but they can handle arbitrary topologies. The methods for supervisory control under partial controllability with regular languages have a polynomial complexity, due to a simplified topology. Comparing the two approaches for supervisory control with full controllability and observability, we notice that both approaches yield necessary and sufficient conditions for the existence of a controller; with the supervisory control approach a single (smallest) controller is derived; with the language approach, the set of all controllers is derived, including the smallest and the largest ones. Both approaches have the same complexity. We also contrast the two approaches on supervisory control with partial controllability and finally illustrate our findings on an example of supervising a cat and mouse in a maze.

Chapter 16 addresses the problem of testing a component FSM in an FSM network, also known as the problem of testing in context or embedded testing. It is customary to assume that a context FSM implementation (the fixed part) needs no testing and conforms to its specification, while an implementation of the embedded FSM (the unknown part) has to be tested for its conformance to the specification FSM. Inputs and outputs of the embedded machine are neither controllable nor observable and tests should be expressed using external inputs and outputs of the context machine. Testing an FSM in isolation can, under certain assumptions, detect faults which can make an implementation non-equivalent to its (deterministic) specification. However, testing in context can be more permissive, since the context FSM can tolerate faults in the embedded machine. We demonstrate first that the largest set of its permissible behaviors obtained by finding the largest solution of an appropriate FSM equation provides the exact characterization of all such faults. Since the largest solution is represented by a non-deterministic FSM such that each non-conforming implementation is not its reduction, a test suite has to be derived using the reduction relation between the implementation and specification FSMs. The existing methods developed for testing FSMs in isolation can be used for this purpose. However, the resulting tests use the internal alphabets of the embedded machine; thus the internal tests must be translated into external tests defined over the external alphabets of the system. The problem of translating internal tests into external ones is sometimes called the fault propagation problem and can also be formulated as solving an appropriate equation over FSMs or automata. We use a simple example to illustrate the use of language equations for solving testing in context.

Chapter 17 discusses the application of language solving to games. We demonstrate how finding winning strategies of some combinatorial games, such as the NIM game, tic-tae-toe, etc., can be formulated as solving the unknown component problem. Therefore, BALM can be used to synthesize winning strategies of these combinatorial games. The approach we take is to describe the dynamics and the state of the game in the fixed component. The unknown component represents one of the players of a two person game. A winning strategy is the ability to make a move to retain the possibility of winning. This is achieved when a solution is prefix-closed and progressive. If the input to the unknown component is the state of the game, then progressive means that, for every state that the game can get into, there is always an input which leads to an accepting state. We illustrate our approach by describing and solving the game of NIM.

Chapter 18 discusses the extension of BALM to synchronous equations over co-Büchi specifications (with a brief mention of Büchi specifications). This extension is motivated by sequential synthesis problems where the objective is to find a strategy, implementable as an FSM, in order to guide a system to a given subset of accepting states with some desirable property and keep it there. Such requirements can be expressed by co-Büchi automata. We propose a synthesis flow, similar to that described in Chap. 2 for regular (finite-word) automata with the main difference in the last step, where the most general solution is trimmed to obtain FSM solutions that meet the co-Büchi condition. This necessitates a special non-regular method, based on formulating the problem as a SAT instance where each satisfying assignment corresponds to a way of trimming the graph to enforce the desired property.

Part I
Theory of Equations Over Languages and Automata

In this part we present the basics of equations over languages and automata, and specialize the theory to finite and ω-regular languages.

In Chap. 2 we define and study *abstract equations over languages*, to obtain results valid for any language equation. We investigate two composition operators for abstract languages: synchronous composition, $\bullet$, and parallel composition, $\diamond$, and we check conformity by language containment.

A key contribution is the *computation of the most general solutions of the language equations* $A \bullet X \subseteq C$ and $A \diamond X \subseteq C$, found respectively as $S = \overline{A \bullet \overline{C}}$, and $S = \overline{A \diamond \overline{C}}$. The derivation sheds lights on the properties required of a composition operator to yield such a closed formula as the largest solution, and explains when different equations give rise to that same type of solution formula. These formulas turn out to subsume a panoply of specialized solutions derived in the past for specific composition operators and topologies. Some common network topologies are shown in Fig. 1.1.

Then in Chap. 3 we specialize language equations to languages associated with classes of automata used for modeling hardware and software systems, namely, regular languages as counterparts of finite automata, FSM languages as counterparts of FSMs. Thus we can operate algorithmically on those languages through their automata and study how to solve effectively their related language equations. It is important to find solutions within the same language class of the equation, e.g., when studying FSM language equations we look for solutions that are FSM languages. Moreover, we are interested in subsets of solutions characterized by further properties of practical interest, e.g., FSM languages that satisfy the Moore property; thus the solutions are restricted further.

In Chap. 4 we study the extensions to equations over ω-automata of the results obtained in the finite regular case.

Various contributions, investigating partial aspects of the topic of this research, have been published. A complete survey is provided in Chap. 5.

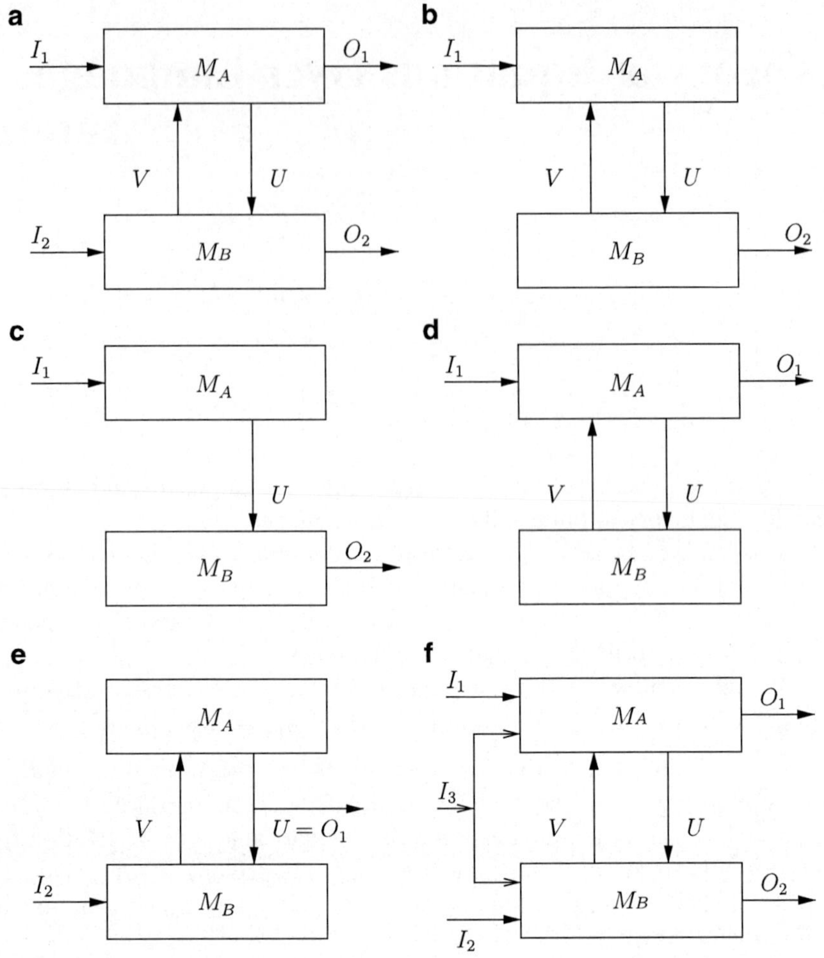

Fig. 1.1 Patterns of composition. (**a**) general topology; (**b**) 2-way cascade (or 2-way series) topology; (**c**) 1-way cascade (or 1-way series) topology; (**d**) rectification topology; (**e**) controller's (or supervisory control) topology; (**f**) variant of general topology

Chapter 2
Equations Over Languages and Finite Automata

2.1 Preliminaries on Languages and Finite Automata

2.1.1 Languages and Operators

Definition 2.1.1. An alphabet is a finite set of symbols. The set of all finite strings over a fixed alphabet X is denoted by $X^\star$. $X^\star$ includes the empty string ϵ. A subset $L \subseteq X^\star$ is called a **language** over alphabet X.

A language L over the alphabet $X \times V$ is a language over the alphabet $A = \{(x, v) \mid x \in X, v \in V\}$. Standard set-theoretic operations are defined on alphabets, e.g., union and intersection.

Some standard operations on languages are:

1. Given languages L_1 and L_2, respectively over alphabets X_1 and X_2, the language $L_1 \cup L_2$ over alphabet $X_1 \cup X_2$ is the **union** of languages L_1 and L_2.
2. Given languages L_1 and L_2, respectively over alphabets X_1 and X_2, the language $L_1 L_2 = \{\alpha\beta \mid \alpha \in L_1, \beta \in L_2\}$ over alphabet $X_1 \cup X_2$ is the **concatenation** of languages L_1 and L_2. Define $L^0 = \{\epsilon\}$, $L^i = LL^{i-1}$. The **Kleene closure** of L is the set $L^\star = \cup_{i=0}^{\infty} L^i$ and the **positive Kleene closure** of L is $L^+ = \cup_{i=1}^{\infty} L^i$. Finally, the l**-bounded Kleene closure** of L is set $L^{\leq l} = \cup_{i=0}^{l} L^i$.
3. Given languages L_1 and L_2, respectively over alphabets X_1 and X_2, the language $L_1 \cap L_2$ over alphabet $X_1 \cap X_2$ is the **intersection** of languages L_1 and L_2. If $X_1 \cap X_2 = \emptyset$ then $L_1 \cap L_2 = \emptyset$.
4. Given a language L over alphabet X, the language $\overline{L} = X^\star \setminus L$ over alphabet X is the **complement** of language L. Similarly, given languages L_1 and L_2, respectively over alphabets X_1 and X_2, the language $L_1 \setminus L_2 = L_1 \cap \overline{L_2}$ over alphabet X_1 is the **difference** of languages L_1 and L_2.
5. Given a language L over alphabet X, the language $Pref(L) = \{x \in X^\star \mid \exists y \in X^\star, xy \in L\}$ is the **prefix-closure** of L, i.e., the language whose words are all the prefixes of words in L.

T. Villa et al., *The Unknown Component Problem: Theory and Applications*,
DOI 10.1007/978-0-387-68759-9_2, © Springer Science+Business Media, LLC 2012

6. Given a language L over alphabet X, L^{Pref} is the **largest prefix-closed language** L' with $L' \subseteq L$.

It is useful to recall the notions of substitution and homomorphism of languages [63]. A **substitution** f is a mapping of an alphabet X onto subsets of $Y^\star$ for some alphabet Y. The substitution f is extended to strings by setting $f(\epsilon) = \{\epsilon\}$ and $f(xa) = f(x)f(a)$. An **homomorphism** h is a substitution such that $h(a)$ is a singleton string for each symbol a in the alphabet X. We introduce some useful operations on languages. The first two are associated with synchronous composition, while the last two are associated with parallel composition. These operations are integral parts of constructing the most general solution.

1. Given a language L over alphabet $X \times V$, consider the homomorphism $p : X \times V \to V^\star$ defined as
$$p((x, v)) = v,$$
then the language
$$L_{\downarrow V} = \{p(\alpha) \mid \alpha \in L\}$$
over alphabet V is the **projection** of language L to alphabet V, or V-projection of L. By definition of substitution $p(\epsilon) = \epsilon$.
2. Given a language L over alphabet X and an alphabet V, consider the substitution $l : X \to 2^{(X \times V)^\star}$ defined as
$$l(x) = \{(x, v) \mid v \in V\},$$
then the language
$$L_{\uparrow V} = \{l(\alpha) \mid \alpha \in L\}$$
over alphabet $X \times V$ is the **lifting** of language L to alphabet V, or V-lifting of L. By definition of substitution $l(\epsilon) = \{\epsilon\}$.
3. Given a language L over alphabet $X \cup V$, consider the homomorphism $r : X \cup V \to V^\star$ defined as
$$r(y) = \begin{cases} y \text{ if } y \in V \\ \epsilon \text{ if } y \in X \setminus V \end{cases},$$
then the language
$$L_{\Downarrow V} = \{r(\alpha) \mid \alpha \in L\}$$
over alphabet V is the **restriction** of language L to alphabet V, or V-restriction of L, i.e., words in $L_{\Downarrow V}$ are obtained from those in L by deleting all the symbols in X that are not in V. By definition of substitution $r(\epsilon) = \epsilon$.
4. Given a language L over alphabet X and an alphabet V, consider the mapping $e : X \to 2^{(X \cup V)^\star}$ defined as
$$e(x) = \{\alpha x \beta \mid \alpha, \beta \in (V \setminus X)^\star\},$$

then the language

$$L_{\Uparrow V} = \{e(\alpha) \mid \alpha \in L\}$$

over alphabet $X \cup V$ is the **expansion** of language L to alphabet V, or V-expansion of L, i.e., words in $L_{\Uparrow V}$ are obtained from those in L by inserting anywhere in them words from $(V \setminus X)^{\star}$. Notice that e is not a substitution and that $e(\epsilon) = \{\alpha \mid \alpha \in (V \setminus X)^{\star}\}$.

Given a language L over alphabet X, an alphabet V, and a natural number l, consider the mapping $e_l : X \to 2^{(X \cup V)^l}$ defined as

$$e_l(x) = \{\alpha x \beta \mid \alpha, \beta \in (V \setminus X)^{\leq l}\},$$

then the language

$$L_{\Uparrow (V,l)} = \{e_l(\alpha) \mid \alpha \in L\}$$

over alphabet $X \cup V$ is the ***l*-bounded expansion** of language L over alphabet V, or (V,l)-expansion of L, i.e., words in $L_{\Uparrow V}$ are obtained from those in L by inserting anywhere in them words from $(V \setminus X)^{\leq l}$. Notice that e_l is not a substitution and that $e_l(\epsilon) = \{\alpha \mid \alpha \in (V \setminus X)^{\leq l}\}$.

By definition $\emptyset_{\downarrow V} = \emptyset$, $\emptyset_{\uparrow V} = \emptyset$, $\emptyset_{\Downarrow V} = \emptyset$, $\emptyset_{\Uparrow V} = \emptyset$, $\emptyset_{\Uparrow (V,l)} = \emptyset$.

The four previous operators change a language and its alphabet of definition; in particular the operators $\uparrow$ and $\downarrow$ change what components are present in the Cartesian product that defines the language alphabet. We assume that each component has a fixed position in the Cartesian product. For instance, let language L_1 be defined over alphabet I and language L_2 be defined over alphabet O, then language $L_{1\uparrow O}$ is defined over alphabet $I \times O$ and also language $L_{2\uparrow I}$ is defined over alphabet $I \times O$, if by assumption I precedes O in the Cartesian product. More precisely, say that we introduce an ordering of alphabets, i, by which I is mapped to index $i(I)$ and O is mapped to $i(O)$, then $i(I) < i(O)$ implies that I precedes O in any Cartesian product of alphabets. The ordering is arbitrary, but, once chosen, it holds through the sequence of language operations.

The following straightforward facts hold between the projection and lifting operators, and between the restriction and expansion operators.

Proposition 2.1. *The following inverse laws for* $\uparrow, \downarrow$ *and* $\Uparrow, \Downarrow$ *hold.*

(a) Let X *and* Y *be alphabets, and let* L *be a language over alphabet* X*, then* $(L_{\uparrow Y})_{\downarrow X} = L$.
(b) Let X *and* Y *be alphabets, and let* L *be a language over alphabet* $X \times Y$*, then* $(L_{\downarrow X})_{\uparrow Y} \supseteq L$.
(c) Let X *and* Y *be disjoint alphabets, and let* L *be a language over alphabet* X*, then* $(L_{\Uparrow Y})_{\Downarrow X} = L$.
(d) Let X *and* Y *be disjoint alphabets, and let* L *be a language over alphabet* $X \cup Y$*, then* $(L_{\Downarrow X})_{\Uparrow Y} \supseteq L$.

Proposition 2.2. *The following equivalences hold.*

(a) Given alphabets X and Y, a language L over alphabet X, and a string $\alpha \in (X \times Y)^\star$, then $\alpha_{\downarrow X} \in L \Leftrightarrow \alpha \in L_{\uparrow Y}$.
(b) Given disjoint alphabets X and Y, a language L over alphabet X, and a string $\alpha \in (X \cup Y)^\star$, then $\alpha_{\Downarrow X} \in L \Leftrightarrow \alpha \in L_{\Uparrow Y}$.

Proposition 2.3. *The following distributive laws for $\uparrow$ and $\downarrow$ hold.*

(a) Let L_1, L_2 be languages over alphabet U. Then $\uparrow$ commutes with $\cup$

$$(L_1 \cup L_2)_{\uparrow I} = L_{1\,\uparrow I} \cup L_{2\,\uparrow I}.$$

(b) Let L_1, L_2 be languages over alphabet U. Then $\uparrow$ commutes with $\cap$

$$(L_1 \cap L_2)_{\uparrow I} = L_{1\,\uparrow I} \cap L_{2\,\uparrow I}.$$

(c) Let M_1, M_2 be languages over alphabet $I \times U$. Then $\downarrow$ commutes with $\cup$

$$(M_1 \cup M_2)_{\downarrow U} = M_{1\,\downarrow U} \cup M_{2\,\downarrow U}.$$

(d) Let M_1, M_2 be languages over alphabet $I \times U$. If $M_2 = (M_{2\,\downarrow U})_{\uparrow I}$ (or $M_1 = (M_{1\,\downarrow U})_{\uparrow I}$), then $\downarrow$ commutes with $\cap$

$$(M_1 \cap M_2)_{\downarrow U} = M_{1\,\downarrow U} \cap M_{2\,\downarrow U}.$$

Proof. Thesis: $(L_1 \cap L_2)_{\uparrow I} = L_{1\,\uparrow I} \cap L_{2\,\uparrow I}$.

($\Rightarrow$) If the string $(i_1, u_1) \ldots (i_k, u_k) \in (L_1 \cap L_2)_{\uparrow I}$, then $u_1 \ldots u_k \in L_1 \cap L_2$; thus $u_1 \ldots u_k \in L_1$, $u_1 \ldots u_k \in L_2$, and so $(i_1, u_1) \ldots (i_k, u_k) \in L_{1\,\uparrow I}$, $(i_1, u_1) \ldots (i_k, u_k) \in L_{2\,\uparrow I}$, implying $(i_1, u_1) \ldots (i_k, u_k) \in L_{1\,\uparrow I} \cap L_{2\,\uparrow I}$.

($\Leftarrow$) If the string $(i_1, u_1) \ldots (i_k, u_k) \in L_{1\,\uparrow I} \cap L_{2\,\uparrow I}$, then $(i_1, u_1) \ldots (i_k, u_k) \in L_{1\,\uparrow I}$, $(i_1, u_1) \ldots (i_k, u_k) \in L_{2\,\uparrow I}$; thus $u_1 \ldots u_k \in L_1$, $u_1 \ldots u_k \in L_2$, implying $u_1 \ldots u_k \in L_1 \cap L_2$, and so $(i_1, u_1) \ldots (i_k, u_k) \in (L_1 \cap L_2)_{\uparrow I}$.

Similarly one proves the first and third identity involving $\cup$.

Thesis: If $M_2 = (M_{2\,\downarrow U})_{\uparrow I}$ (or $M_1 = (M_{1\,\downarrow U})_{\uparrow I}$), then $(M_1 \cap M_2)_{\downarrow U} = M_{1\,\downarrow U} \cap M_{2\,\downarrow U}$.

($\Rightarrow$) If the string $u_1 \ldots u_k \in (M_1 \cap M_2)_{\downarrow U}$ then there exists $i_1 \ldots i_k$ such that $(i_1, u_1) \ldots (i_k, u_k) \in M_1 \cap M_2$, i.e., $(i_1, u_1) \ldots (i_k, u_k) \in M_1$, $(i_1, u_1) \ldots (i_k, u_k) \in M_2$, and so $u_1 \ldots u_k \in M_{1\,\downarrow U}$ and $u_1 \ldots u_k \in M_{2\,\downarrow U}$.

($\Leftarrow$) If the string $u_1 \ldots u_k \in M_{1\,\downarrow U} \cap M_{2\,\downarrow U}$, i.e., $u_1 \ldots u_k \in M_{1\,\downarrow U}$ and $u_1 \ldots u_k \in M_{2\,\downarrow U}$, then there exists $i_1 \ldots i_k$ such that $(i_1, u_1) \ldots (i_k, u_k) \in M_1$. Moreover, since $M_2 = (M_{2\,\downarrow U})_{\uparrow I}$, from $u_1 \ldots u_k \in M_{2\,\downarrow U}$ it follows that $(i_1, u_1) \ldots (i_k, u_k) \in M_2$. In summary, $(i_1, u_1) \ldots (i_k, u_k) \in M_1$ and $(i_1, u_1) \ldots (i_k, u_k) \in M_2$, implying $(i_1, u_1) \ldots (i_k, u_k) \in M_1 \cap M_2$, from which follows $u_1 \ldots u_k \in (M_1 \cap M_2)_{\downarrow U}$. □

Corollary 2.4. *The following commutative laws for* $\uparrow$ *and* $\downarrow$ *hold.*

(a) Let $L_i, i = 1, \ldots, n$*, be languages over alphabet* U*. Then* $\uparrow$ *commutes with both* $\cup$ *and* $\cap$

$$(\cup L_i)_{\uparrow I} = \cup(L_i \,{}_{\uparrow I}),$$

$$(\cap L_i)_{\uparrow I} = \cap(L_i \,{}_{\uparrow I}).$$

(b) Let $M_i, i = 1, \ldots, n$*, be languages over alphabet* $I \times U$*. Then* $\downarrow$ *commutes with* $\cup$

$$(\cup M_i)_{\downarrow U} = \cup(M_i \,{}_{\downarrow U}).$$

(c) Let $M_i, i = 1, \ldots, n$*, be languages over alphabet* $I \times U$*. If* $M_2 = (M_2 \,{}_{\downarrow U})_{\uparrow I}$*, ...,* $M_n = (M_n \,{}_{\downarrow U})_{\uparrow I}$ *(or any collection of* $n - 1$ *languages* M_i *satisfies this property), then* $\downarrow$ *commutes with* $\cap$

$$(\cap M_i)_{\downarrow U} = \cap(M_i \,{}_{\downarrow U}).$$

The proof is by induction based on Prop. 2.3.

Proposition 2.5. *Suppose that* I *and* U *are disjoint alphabets. The following distributive laws for* $\Uparrow$ *and* $\Downarrow$ *hold.*

(a) Let L_1, L_2 *be languages over alphabet* U*. Then* $\Uparrow$ *commutes with* $\cup$

$$(L_1 \cup L_2)_{\Uparrow I} = L_1 \,{}_{\Uparrow I} \cup L_2 \,{}_{\Uparrow I}.$$

(b) Let L_1, L_2 *be languages over alphabet* U*. Then* $\Uparrow$ *commutes with* $\cap$

$$(L_1 \cap L_2)_{\Uparrow I} = L_1 \,{}_{\Uparrow I} \cap L_2 \,{}_{\Uparrow I}.$$

(c) Let M_1, M_2 *be languages over alphabet* $I \cup U$*. Then* $\Downarrow$ *commutes with* $\cup$

$$(M_1 \cup M_2)_{\Downarrow U} = M_1 \,{}_{\Downarrow U} \cup M_2 \,{}_{\Downarrow U}.$$

(d) Let M_1, M_2 *be languages over alphabet* $I \cup U$*. If* $M_2 = (M_2 \,{}_{\Downarrow U})_{\Uparrow I}$ *(or* $M_1 = (M_1 \,{}_{\Downarrow U})_{\Uparrow I}$*) then* $\Downarrow$ *commutes with* $\cap$

$$(M_1 \cap M_2)_{\Downarrow U} = M_1 \,{}_{\Downarrow U} \cap M_2 \,{}_{\Downarrow U}.$$

Proof. Thesis: $(L_1 \cap L_2)_{\Uparrow I} = L_1 \,{}_{\Uparrow I} \cap L_2 \,{}_{\Uparrow I}$.

($\Rightarrow$) If the string $\alpha_1 u_1 \ldots \alpha_k u_k \alpha_{k+1} \in (L_1 \cap L_2)_{\Uparrow I}$ and $\alpha_1, \ldots, \alpha_k, \alpha_{k+1} \in I^\star$, then $u_1 \ldots u_k \in L_1 \cap L_2$; thus $u_1 \ldots u_k \in L_1$, $u_1 \ldots u_k \in L_2$, and so $\alpha_1 u_1 \ldots \alpha_k u_k \alpha_{k+1} \in L_1 \,{}_{\Uparrow I}$, $\alpha_1 u_1 \ldots \alpha_k u_k \alpha_{k+1} \in L_2 \,{}_{\Uparrow I}$, implying $\alpha_1 u_1 \ldots \alpha_k u_k \alpha_{k+1} \in L_1 \,{}_{\Uparrow I} \cap L_2 \,{}_{\Uparrow I}$.

($\Leftarrow$) If the string $\alpha_1 u_1 \dots \alpha_k u_k \alpha_{k+1} \in L_{1\Uparrow I} \cap L_{2\Uparrow I}$, then it holds that $\alpha_1 u_1 \dots \alpha_k u_k \alpha_{k+1} \in L_{1\Uparrow I}$ and $\alpha_1 u_1 \dots \alpha_k u_k \alpha_{k+1} \in L_{2\Uparrow I}$; thus $u_1 \dots u_k \in L_1$, $u_1 \dots u_k \in L_2$, implying $u_1 \dots u_k \in L_1 \cap L_2$, and so it is also $\alpha_1 u_1 \dots \alpha_k u_k \alpha_{k+1} \in (L_1 \cap L_2)_{\Uparrow I}$.

Similarly one proves the first and third identity involving $\cup$.

Thesis: if $M_2 = (M_{2\Downarrow U})_{\Uparrow I}$ (or $M_1 = (M_{1\Downarrow U})_{\Uparrow I}$) then $(M_1 \cap M_2)_{\Downarrow U} = M_{1\Downarrow U} \cap M_{2\Downarrow U}$.

($\Rightarrow$) If the string $u_1 \dots u_k \in (M_1 \cap M_2)_{\Downarrow U}$ then there exists $\alpha_1, \dots \alpha_k, \alpha_{k+1} \in I^\star$ such that it holds that the string $\alpha_1 u_1 \dots \alpha_k u_k \alpha_{k+1} \in M_1 \cap M_2$, i.e., $\alpha_1 u_1 \dots \alpha_k u_k \alpha_{k+1} \in M_1$, $\alpha_1 u_1 \dots \alpha_k u_k \alpha_{k+1} \in M_2$, and so $u_1 \dots u_k \in M_{1\Downarrow U}$ and $u_1 \dots u_k \in M_{2\Downarrow U}$.

($\Leftarrow$) If the string $u_1 \dots u_k \in M_{1\Downarrow U} \cap M_{2\Downarrow U}$, i.e., $u_1 \dots u_k \in M_{1\Downarrow U}$ and $u_1 \dots u_k \in M_{2\Downarrow U}$, then there exists $\alpha_1 \dots \alpha_k \alpha_{k+1} \in I^\star$ such that $\alpha_1 u_1 \dots \alpha_k u_k \alpha_{k+1} \in M_1$. Moreover, since $M_2 = (M_{2\Downarrow U})_{\Uparrow I}$, from $u_1 \dots u_k \in M_{2\Downarrow U}$ it follows that $\alpha_1 u_1 \dots \alpha_k u_k \alpha_{k+1} \in M_2$. In summary, $\alpha_1 u_1 \dots \alpha_k u_k \alpha_{k+1} \in M_1$ and $\alpha_1 u_1 \dots \alpha_k u_k \alpha_{k+1} \in M_2$, implying $\alpha_1 u_1 \dots \alpha_k u_k \alpha_{k+1} \in M_1 \cap M_2$, from which follows $u_1 \dots u_k \in (M_1 \cap M_2)_{\Downarrow U}$. □

Example 2.6. The identity $(M_1 \cap M_2)_{\Downarrow U} = M_{1\Downarrow U} \cap M_{2\Downarrow U}$ does not hold without the additional hypothesis in Prop. 2.5(d). Consider $I = \{a, b\}$, $U = \{u\}$, $M_1 = \{au\}$, $M_2 = \{bu\}$, then $(M_1 \cap M_2)_{\Downarrow U} = \emptyset_{\Downarrow U} = \emptyset$ and $M_{1\Downarrow U} \cap M_{2\Downarrow U} = \{u\} \cap \{u\} = \{u\}$. Notice that au and bu are words of length 2 on the alphabet $I \cup U$.

Proposition 2.7. *The following equivalences hold.*

(a) Let L be a language over alphabet I, then $L_{\uparrow O} = \emptyset \Leftrightarrow L = \emptyset$
(b) Let L be a language over alphabet $I \times O$, then $L_{\downarrow O} = \emptyset \Leftrightarrow L = \emptyset$ In the next two statements, suppose that I and O are disjoint alphabets.
(c) Let L be a language over alphabet $I \cup O$, then $L_{\Uparrow O} = \emptyset \Leftrightarrow L = \emptyset$
(d) Let L be a language over alphabet $I \cup O$, then $L_{\Downarrow O} = \emptyset \Leftrightarrow L = \emptyset$.

Proof. The proofs are straightforward; implication $\Rightarrow$ of statement (d) is true because, even in the case that all strings in L are defined only over symbols from alphabet I, their restriction to alphabet O yields the empty string ϵ (i.e., $\epsilon \in L_{\Downarrow O} \neq \emptyset$) and so from $L \neq \emptyset$ follows that $L_{\Downarrow O} \neq \emptyset$. □

2.1.2 Finite Automata and Regular Expressions

Definition 2.1.2. A **finite automaton** (FA) is a 5-tuple $F = \langle S, \Sigma, \Delta, r, Q \rangle$. S represents the finite state space, Σ represents the finite alphabet, and $\Delta \subseteq \Sigma \times S \times S$ is the next state relation, such that $(i, p, n) \in \Delta$ iff $n \in S$ is a next state of present state $p \in S$ on symbol $i \in \Sigma$. The initial or reset state is $r \in S$ and $Q \subseteq S$ is the set of final or accepting states. A variant of FAs allows the introduction of ϵ-moves, meaning that $\Delta \subseteq (\Sigma \cup \{\epsilon\}) \times S \times S$.

The next state relation Δ can be extended to have as argument strings in Σ^* (i.e., $\tilde{\Delta} \subseteq \Sigma^* \times S \times S$) as follows: $(\rho i, s, s'') \in \tilde{\Delta}$ iff there exists $s' \in S$ such that $(\rho, s, s') \in \tilde{\Delta}$ and $(i, s', s'') \in \Delta$. For ease of notation we just use the symbol Δ for both relations and let the context determine the meaning.

A string x is said to be **accepted** by the FA F if there exists a sequence of transitions corresponding to x such that there exists a state $r' \in Q$ for which $\Delta(x, r, r')$. The **language** accepted by F, designated $L_r(F)$, is the set of strings $\{x \mid \exists r' \in Q\ [\Delta(x, r, r')]\}$. The language accepted or **recognized** by $s \in S$, denoted $L_r(F|s)$ or $L_r(s)$ when F is clear from the context, is the set of strings $\{x \mid \Delta(x, r, s)\}$.

If for each present state p and symbol i there is at least one next state n such that $(i, p, n) \in \Delta$, the FA is said to be **complete**, otherwise it is **partial**. A partial automaton can be made complete by directing the unspecified transitions to a non-accepting state.

An FA is a **deterministic finite automaton** (DFA) if for each present state p and symbol i there is exactly one next state n such that $(i, p, n) \in \Delta$.[1] The relation Δ can be replaced by the next state function δ, defined as $\delta : \Sigma \times S \to S$, where $n \in S$ is the next state of present state $p \in S$ on symbol $i \in \Sigma$ iff $n = \delta(i, p)$. An FA that is not a DFA is a **non-deterministic finite automaton** (NDFA).

A string x is said to be **accepted** by the DFA F if $\delta(x, r) \in Q$. The **language** accepted by F, designated $L_r(F)$, is the set of strings $\{x \mid \delta(x, r) \in Q\}$. The language accepted or **recognized** by $s \in S$, denoted $L_r(F|s)$ or $L_r(s)$ when F is clear from the context, is the set of strings $\{x \mid \delta(x, r) = s\}$.

The languages associated with finite automata are the regular languages, defined by means of regular expressions, as established by the theorem due to Kleene [63].

Definition 2.1.3. The **regular expressions** over an alphabet Σ are defined recursively as follows:

1. $\emptyset$ is a regular expression and denotes the empty set.
2. ϵ is a regular expression and denotes the set $\{\epsilon\}$.
3. For each $a \in \Sigma$, a is a regular expression and denotes the set $\{a\}$.
4. If l_1 and l_2 are regular expressions denoting the languages L_1 and L_2, respectively, then $(l_1 + l_2)$, $(l_1 l_2)$ and $(l_1^\star)$ are regular expressions that denote the sets $L_1 \cup L_2$, $L_1 L_2$ and $L_1^\star$, respectively.

The sets denoted by regular expressions are the **regular languages**.

Regular languages are closed under union, concatenation, complementation and intersection. Also regular languages are closed under projection, lifting and restriction, because they are closed under substitution [63]. Regular languages are closed under expansion, as shown in Sect. 2.3.1 providing an algorithm that, given the finite automaton of a language, returns the finite automaton of the expanded language.

---

[1] By the given definition in agreement with standard textbooks (see [63]), a DFA must be a complete FA, but a complete FA does not need to be deterministic.

2.1.3 Classes of Languages

We introduce several classes of languages used later in the paper.

Definition 2.1.4. A language L over alphabet X is **prefix-closed** if $\forall \alpha \in X^\star \ \forall x \in X \ [\alpha x \in L \Rightarrow \alpha \in L]$. Equivalently, L is prefix-closed iff $L = Pref(L)$.

Definition 2.1.5. A language L over alphabet $X = I \times O$ is **I-progressive** if

$$\forall i \in I \ \exists o \in O \ [\alpha \in L \Rightarrow \alpha \ (i, o) \in L].$$

If L is not I-progressive, then $Prog(L)$ is the **largest I-progressive language** L' such that $L' \subseteq L$.

Definition 2.1.6. A language L over alphabet $I \times O$ is **$I_{\downarrow}$-defined** if $L_{\downarrow I} = I^\star$.

If a language over $X = I \times O$ is I-progressive it is also $I_{\downarrow}$-defined, but the converse does not hold.

Example 2.8. The language $L = \{\epsilon + i_1 o_1 + i_1 o_2 (i_1 o_1)^\star\}$ is $I_{\downarrow}$-defined, but not I-progressive, as witnessed by $\alpha = i_1 o_1 \in L$ and $i = i_1$ for which there is no o such that $\alpha \ i_1 o \in L$.

Definition 2.1.7. A language L over alphabet $X = I \times O$ is **Moore**[2] with respect to alphabet I, if

$$\forall \alpha \in L \ \forall (i, o) \in X \ \forall (i', o') \in X \ [\alpha \ (i, o) \in L \Rightarrow [\alpha \ (i', o') \in L \Rightarrow \alpha \ (i', o) \in L]].$$

Definition 2.1.8. A language $L \subseteq (IO)^\star$ over alphabet $I \cup O$ (I and O disjoint) is ***IO*-prefix-closed** if

$$\forall \alpha \in (IO)^\star \ \forall io \in IO \ [\alpha \ io \in L \Rightarrow \alpha \in L].$$

Definition 2.1.9. A language $L \subseteq (IO)^\star$ over alphabet $I \cup O$ (I and O disjoint) is ***IO*-progressive** if

$$\forall i \in I \ \exists o \in O \ [\alpha \in L \Rightarrow \alpha \ io \in L].$$

Definition 2.1.10. A language $L \subseteq (IU^\star O)^\star$ over alphabet $I \cup U \cup O$ (I, U and O pairwise disjoint) is **$IU^\star O$-progressive** if

$$\forall i \in I \ \exists \beta \in U^\star \ \exists o \in O \ [\alpha \in L \Rightarrow \alpha \ i \beta o \in L].$$

---

[2]This definition is an abstraction to languages of the most common definition of Moore automata/finite state machines.

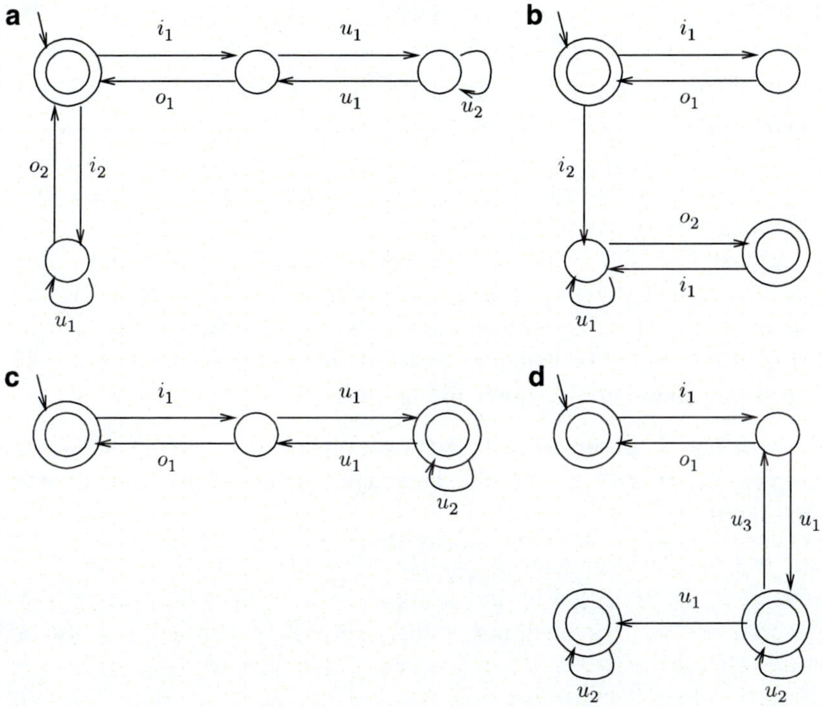

Fig. 2.1 (**a**) Finite automaton of the language described in Example 2.9-a; (**b**) Finite automaton of the language described in Example 2.9-b; (**c**) Finite automaton of the language described in Example 2.10-a; (**d**) Finite automaton of the language described in Example 2.10-b

Example 2.9. (a) Let $I = \{i_1, i_2\}$, $O = \{o_1, o_2\}$ and $U = \{u_1, u_2\}$. The language $L = \{(i_1 u_1 u_2^\star u_1 o_1 + i_2 u_1^\star o_2)^\star\}$ is $UI^\star O$-progressive, since any word in L can be extended to a word in L by suffixes starting with either i_1 or i_2. The corresponding automaton is shown in Fig. 2.1a.

(b) Let $I = \{i_1, i_2\}$, $O = \{o_1, o_2\}$ and $U = \{u_1\}$. The language $L = \{(i_1 o_1)^\star + (i_1 o_1)^\star i_2 u_1^\star o_2 (i_1 u_1^\star o_2)^\star\}$ is not $IU^\star O$-progressive, since the words in the set $\{i_2 u_1^\star o_2 (i_1 u_1^\star o_2)^\star\}$ are in L, but when $i = i_2$ there is no $\beta \in U^\star$ and no $o \in O$ such that $\alpha\, i_2 \beta o \in L$ (e.g., $\alpha = i_2 u_1 o_2$ cannot be extended by any suffix starting with i_2). The corresponding automaton is shown in Fig. 2.1b.

Definition 2.1.11. A language L over alphabet $I \cup O$ (I and O disjoint) is $I_{\Downarrow}$**-defined** if $L_{\Downarrow I} = I^\star$.

An IO-progressive language is $I_{\Downarrow}$-defined, so is an $IU^\star O$-progressive language, but the converse does not hold.

Definition 2.1.12. A language $L \neq \emptyset$ over alphabet $X \cup U$ (X and U disjoint) is U**-deadlock-free** if

$$\forall \alpha \in (X \cup U)^\star\ \forall u \in U\ \exists \beta \in U^\star\ \exists x \in X\ [\alpha u \in L \Rightarrow \alpha u\, \beta x \in L].$$

Any language $L \subseteq (IU^\star O)^\star$ is U-deadlock-free (because no word ending by a symbol $u \in U$ belongs to the language).

Example 2.10. (a) Let $X = I \cup O$, $I = \{i_1, i_2\}$, $O = \{o_1, o_2\}$ and $U = \{u_1, u_2\}$. The language $L = \{(i_1(u_1u_2^\star u_1)^\star o_1)^\star + (i_1(u_1u_2^\star u_1)^\star o_1)^\star i_1u_1u_2^\star\}$ is U-deadlock-free, because any word in the language terminating by u_1 or u_2 can be extended by suffix u_1 to a word in the language terminating by o_1. The corresponding automaton is shown in Fig. 2.1c.

(b) Let $X = I \cup O$, $I = \{i_1, i_2\}$, $O = \{o_1, o_2\}$ and $U = \{u_1, u_2, u_3\}$. The language $L = \{i_1(u_1u_2^\star u_3)^\star o_1)^\star + (i_1(u_1u_2^\star u_3)^\star o_1)^\star i_1u_1u_2^\star + (i_1(u_1u_2^\star u_3)^\star o_1)^\star i_1u_1u_2^\star\, u_1u_2^\star\}$ is not U-deadlock-free, since the words in the collection $\{(i_1(u_1u_2^\star u_3)^\star o_1)^\star i_1\, u_1u_2^\star u_1u_2^\star\}$ cannot be extended to words in L (e.g., $\alpha = i_1u_1u_2u_1$). The corresponding automaton is shown in Fig. 2.1d.

Definition 2.1.13. A language $L \neq \emptyset$ over alphabet $X \cup U$ (X and U disjoint) is **U-convergent** if $\forall \alpha \in X^\star$ the language $\alpha_{\Uparrow U} \cap L$ is finite, otherwise it is **U-divergent**.

Example 2.11. The language $L = \{iu^\star o\}$ where $X = \{i, o\}$ and $U = \{u\}$ is U-divergent, as witnessed by the string $\alpha = io \in X$ whose expansion includes the infinite set $\{iu^\star o\}$ coinciding with L: $\{\alpha_{\Uparrow U}\} = \{(io)_{\Uparrow\{u\}}\} = \{u^\star iu^\star ou^\star\} \supset \{iu^\star o\} = L$.

2.1.4 Composition of Languages

Consider two systems A and B with associated languages $L(A)$ and $L(B)$. The systems communicate with each other by a channel U and with the environment by channels I and O. We introduce two composition operators that describe the external behavior of the composition of $L(A)$ and $L(B)$.

Definition 2.1.14. Given the pairwise disjoint alphabets I, U, O, language L_1 over $I \times U$ and language L_2 over $U \times O$, the **synchronous composition** of languages L_1 and L_2 is the language[3] $[(L_1)_{\uparrow O} \cap (L_2)_{\uparrow I}]_{\downarrow I \times O}$, denoted by $L_1 \bullet_{I \times O} L_2$, defined over $I \times O$.

Definition 2.1.15. Given the pairwise disjoint alphabets I, U, O, language L_1 over $I \cup U$ and language L_2 over $U \cup O$, the **parallel composition** of languages L_1 and L_2 is the language $[(L_1)_{\Uparrow O} \cap (L_2)_{\Uparrow I}]_{\Downarrow I \cup O}$, denoted by $L_1 \diamond_{I \cup O} L_2$, defined over $I \cup O$.

Given alphabets I, U, O, language L_1 over $I \cup U$ and language L_2 over $U \cup O$, the **l-bounded parallel composition** of languages L_1 and L_2 is the language $[(L_1)_{\Uparrow O} \cap (L_2)_{\Uparrow I} \cap (I \cup O)^\star_{\Uparrow (U,l)}]_{\Downarrow I \cup O}$, denoted by $L_1 \diamond_{l_{I \cup O}} L_2$, defined over $I \cup O$.

[3]Use the same order $I \times U \times O$ in the languages $(L_1)_{\uparrow O}$ and $(L_2)_{\uparrow I}$.

Fig. 2.2 Composition topology with three components

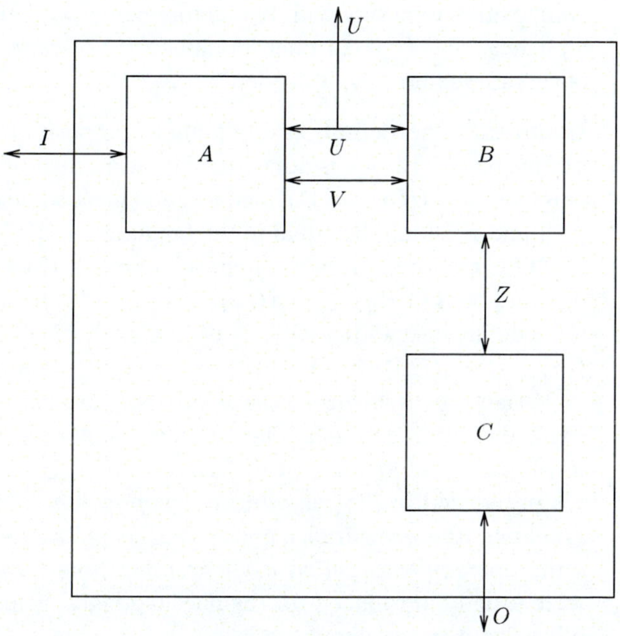

For ease of notation, we will omit the alphabet from the symbol of synchronous and parallel composition, unless unclear from the context. By definition of the operations $_{\downarrow V}$, $_{\uparrow V}$, $_{\Downarrow V}$, $_{\Uparrow V}$, $_{\Uparrow (V,l)}$ it follows that $\emptyset \bullet L = L \bullet \emptyset = \emptyset$, $\emptyset \diamond L = L \diamond \emptyset = \emptyset$, $\emptyset \diamond_l L = L \diamond_l \emptyset = \emptyset$.

When $l = \infty$ the definition of l-bounded parallel composition reduces to the definition of parallel composition of languages, because then $(I \cup O)^{\star}_{\Uparrow (U,l)}$ becomes $(I \cup O \cup U)^{\star}$, that is the universe over $I \cup O \cup U$, and so it can be dropped from the conjunction.

These definitions can be easily extended to more components and more complex interconnection topologies, e.g., to the topology where U is observable externally or where U is the cartesian product of two alphabets only one of which is observable.

For instance in Fig. 2.2 we show a composition topology with three components: A, B and C, which together define the composed system:

$$(A_{\uparrow Z \times O} \cap B_{\uparrow I \times O} \cap C_{\uparrow I \times U \times V})_{\downarrow I \times U \times O}.$$

We notice that I and O are external variables, U is an internal variable that is observable externally, whereas V and Z are internal variables. The composition is well-formed because the synchronous composition operator is associative [150]. Each specific topology dictates the alphabets to which projection and lifting (restriction and expansion) should be applied. In the most straightforward composition topology with two components, we will assume that I, O are external alphabets, and U is an internal alphabet; however, U or a part of it can be made observable externally, if needed.

Comment. The definition of parallel composition justifies *a-posteriori* why the expansions operator e is not defined to be a substitution, i.e., $e(\epsilon) \neq \{\epsilon\}$. Consider a language $A = ((io)^\star(uv)^\star)^\star$ and a language B whatsoever. The parallel composition of A and B should be equal to the language $(io)^\star$, because B should not affect the $I \cup O$ behavior of A. Now suppose $B = \{\epsilon\}$. If we would define $e(\epsilon) = \{\epsilon\}$, then it would be $A \cap B_{\Uparrow I \cup O} = ((io)^\star(uv)^\star)^\star \cap \{\epsilon\}_{\Uparrow I \cup O} = ((io)^\star(uv)^\star)^\star \cap \{\epsilon\} = \{\epsilon\}$; if we define instead $\{\epsilon\}_{\Uparrow I \cup O} = (I \cup O)^\star$ then it is $A \cap B_{\Uparrow I \cup O} = ((io)^\star(uv)^\star)^\star \cap \{\epsilon\}_{\Uparrow I \cup O} = ((io)^\star(uv)^\star)^\star \cap (i \cup o)^\star = (io)^\star$, that is the expected result.

Variants of *synchronous composition* are introduced in [25] as *product*, $\times$ (with the comment *sometimes called completely synchronous composition*), and in [82] as *synchronous parallel composition*, $\otimes$. Variants of *parallel composition* are introduced in [25] as *parallel composition*, $\|$ (with the comment *often called synchronous composition*), and in [82] as *interleaving parallel composition*, $\|$; the same operator was called *asynchronous composition* in [107]. These definitions were usually introduced for regular languages; actually they were more commonly given for finite automata.

It has also been noticed by Kurshan [82] and Arnold [2] that asynchronous systems can also be modeled with the synchronous interpretation, using null transitions to keep a transition system in the same state for an arbitrary period of time. Kurshan [82] observes that: "While synchronous product often is thought to be a simple -even uninteresting!- type of coordination, it can be shown that, through use of nondeterminism, this conceptually simple coordination serves to model the most general 'asynchronous' coordination, i.e., where processes progress at arbitrary rates relative to one another. In fact the 'interleaving' model, the most common model for asynchrony in the software community, can be viewed as a special case of this synchronous product." A technical discussion can be found in [83], but the transformation is not straightforward in practice, and the matter requires further investigation.

In the sequel it will be useful to extend some properties of languages to the composition of two languages. As examples, we illustrate the extension for I-progressive and $I^\star O$-progressive languages.

Definition 2.1.16. Given a language A over alphabet $I \times U$, a language B over alphabet $U \times O$ is **A-compositionally I-progressive** if the language $L = A_{\uparrow O} \cap B_{\uparrow I}$ over alphabet $X = I \times U \times O$ is I-progressive, i.e., $\forall i \in I \; \exists (u,o) \in U \times O \; [\alpha \in L \Rightarrow \alpha \, (i,u,o) \in L]$.

Definition 2.1.17. Given a language A over alphabet $I \cup U$, a language B over alphabet $U \cup O$ is **A-compositionally $IU^\star O$-progressive** if the language $L = A_{\uparrow O} \cap B_{\uparrow I} \subseteq (IU^\star O)^\star$ over alphabet $X = I \cup U \cup O$ (I, U and O pairwise disjoint) is $IU^\star O$-progressive, i.e., $\forall i \in I \; \exists \beta \in U^\star \; \exists o \in O \; [\alpha \in L \Rightarrow \alpha \, i \beta o \in L]$.

Defns. 2.1.16 and 2.1.17 characterize compositions that do not fall into a deadlock or a livelock.

When clear from the context, instead of *A-compositionally* we will write more simply *compositionally*.

2.2 Solution of Equations Over Languages

2.2.1 Language Equations Under Synchronous Composition

Given the alphabets I, U, O, a language A over alphabet $I \times U$ and a language C over alphabet $I \times O$, consider the language equations

$$A \bullet X \subseteq C, \tag{2.1}$$

or,

$$A \bullet X = C. \tag{2.2}$$

Definition 2.2.1. Given the alphabets I, U, O, a language A over alphabet $I \times U$ and a language C over alphabet $I \times O$, language B over alphabet $U \times O$ is called a **solution** of the equation $A \bullet X \subseteq C$ iff $A \bullet B \subseteq C$.

Given the alphabets I, U, O, a language A over alphabet $I \times U$ and a language C over alphabet $I \times O$, language B over alphabet $U \times O$ is called a **solution** of the equation $A \bullet X = C$ iff $A \bullet B = C$.

A solution is called the **largest solution** if it contains any other solution. $B = \emptyset$ is the trivial solution.

Theorem 2.12. *The largest solution of the equation $A \bullet X \subseteq C$ is the language $S = \overline{A \bullet \overline{C}}$.*

If $A \bullet \overline{\overline{A \bullet \overline{C}}} = S$ then $\overline{A \bullet \overline{C}}$ is the largest solution of the equation $A \bullet X = C$.

Proof. Consider a string $\alpha \in (U \times O)^\star$; then α is in the largest solution of $A \bullet X \subseteq C$ iff $A \bullet \{\alpha\} \subseteq C$ and the following chain of equivalences follows:

$$A \bullet \{\alpha\} \subseteq C \Leftrightarrow$$

$$(A_{\uparrow O} \cap \{\alpha\}_{\uparrow I})_{\downarrow I \times O} \cap \overline{C} = \emptyset \Leftrightarrow \textit{by Prop. } 2.1(a)\ \overline{C} = (\overline{C}_{\uparrow U})_{\downarrow I \times O}$$

$$(A_{\uparrow O} \cap \{\alpha\}_{\uparrow I})_{\downarrow I \times O} \cap (\overline{C}_{\uparrow U})_{\downarrow I \times O} = \emptyset \Leftrightarrow \textit{by Prop. } 2.3(d)$$

$$\textit{since } ((\overline{C}_{\uparrow U})_{\downarrow I \times O})_{\uparrow U} = \overline{C}_{\uparrow U}$$

$$(A_{\uparrow O} \cap \{\alpha\}_{\uparrow I} \cap \overline{C}_{\uparrow U})_{\downarrow I \times O} = \emptyset \Leftrightarrow \textit{by Prop. } 2.7(b)$$

$$A_{\uparrow O} \cap \{\alpha\}_{\uparrow I} \cap \overline{C}_{\uparrow U} = \emptyset \Leftrightarrow \textit{by Prop. } 2.7(b)$$

$$(A_{\uparrow O} \cap \{\alpha\}_{\uparrow I} \cap \overline{C}_{\uparrow U})_{\downarrow U \times O} = \emptyset \Leftrightarrow \textit{by Prop. } 2.3(d)$$

$$\textit{since } \{\alpha\}_{\uparrow I} = ((\{\alpha\}_{\uparrow I})_{\downarrow U \times O})_{\uparrow I}$$

$$(\{\alpha\}_{\uparrow I})_{\downarrow U \times O} \cap (A_{\uparrow O} \cap \overline{C}_{\uparrow U})_{\downarrow U \times O} = \emptyset \Leftrightarrow \textit{by Prop. } 2.1(a)\ (\{\alpha\}_{\uparrow I})_{\downarrow U \times O} = \{\alpha\}$$

$$\{\alpha\} \cap (A_{\uparrow O} \cap \overline{C}_{\uparrow U})_{\downarrow U \times O} = \emptyset \Leftrightarrow$$

$$\alpha \notin (A_{\uparrow O} \cap \overline{C}_{\uparrow U})_{\downarrow U \times O} \Leftrightarrow$$

$$\alpha \in \overline{(A_{\uparrow O} \cap \overline{C}_{\uparrow U})_{\downarrow U \times O}} \Leftrightarrow$$

$$\alpha \in \overline{A \bullet \overline{C}}$$

Therefore the largest solution of the language equation $A \bullet X \subseteq C$ is given by the language

$$S = \overline{A \bullet \overline{C}}. \tag{2.3}$$

□

Corollary 2.13. *A language B over alphabet $U \times O$ is a solution of $A \bullet X \subseteq C$ iff $B \subseteq \overline{A \bullet \overline{C}}$.*

Equations over languages can be extended to topologies with more than two components, as the one in Fig. 2.3, whose largest solution is given by:

$$S = \overline{((A_{\uparrow Z \times O} \cap C_{\uparrow I \times U \times V}) \cap \overline{S}_{\uparrow V \times Z})_{\downarrow U \times V \times Z}}$$

(see [150]).

Let S be the largest solution of the equation $A \bullet X \subseteq C$. It is of interest to investigate subsets of S that satisfy some further properties, e.g., being prefix-closed, progressive, etc.

If S is prefix-closed then S is the largest prefix-closed solution of the equation. However, not every non-empty subset of S inherits the feature of being prefix-closed. If S is not prefix-closed, then denote by S^{Pref} the set obtained from S by deleting each string that has a prefix not in S.

Proposition 2.14. *If $S^{Pref} \neq \emptyset$, then S^{Pref} is the largest prefix-closed solution of the equation $A \bullet X \subseteq C$.*

If $S^{Pref} = \emptyset$, then the equation $A \bullet X \subseteq C$ has no prefix-closed solution.

If the language S does not include the empty string, then $A \bullet X \subseteq C$ has no prefix-closed solution.

If S is U-progressive (S is a language over alphabet $U \times O$), then S is the largest U-progressive solution of the equation. However, not each non-empty subset of S

Fig. 2.3 Language equation with respect to a composition topology with three components

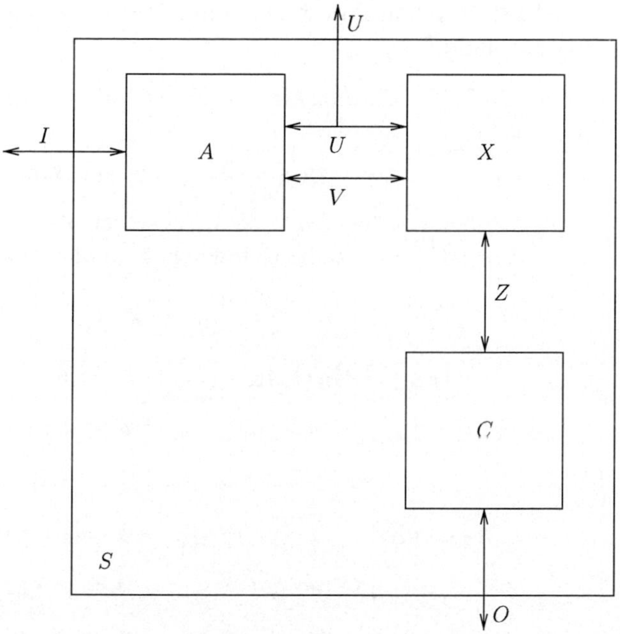

inherits the feature of being U-progressive. If S is not U-progressive, then denote by $Prog(S)$ the set obtained from S by deleting each string α such that, for some $u \in U$, there is no $o \in O$ for which $\alpha(u, o) \in S$.

Proposition 2.15. *If $Prog(S) \neq \emptyset$, then the language $Prog(S)$ is the largest U-progressive solution of the equation $A \bullet X \subseteq C$.*

If $Prog(S) = \emptyset$, then the equation $A \bullet X \subseteq C$ has no U-progressive solution.

2.2.2 Language Equations Under Parallel Composition

Given the pairwise disjoint alphabets I, U, O, a language A over alphabet $I \cup U$, and a language C over alphabet $I \cup O$, consider the language equation

$$A \diamond X \subseteq C, \tag{2.4}$$

or,

$$A \diamond X = C. \tag{2.5}$$

Definition 2.2.2. Given the pairwise disjoint alphabets I, U, O, a language A over alphabet $I \cup U$ and a language C over alphabet $I \cup O$, language B over alphabet $U \cup O$ is called a **solution** of the equation $A \diamond X \subseteq C$ iff $A \diamond B \subseteq C$.

Given the pairwise disjoint alphabets I, U, O, a language A over alphabet $I \cup U$ and a language C over alphabet $I \cup O$, language B over alphabet $U \cup O$ is called a **solution** of the equation $A \diamond X = C$ iff $A \diamond B = C$.

The **largest solution** is a solution that contains any other solution. $B = \emptyset$ is the trivial solution.

Theorem 2.16. *The largest solution of the equation $A \diamond X \subseteq C$ is the language $S = \overline{A \diamond \overline{C}}$.*

If $A \diamond \overline{A \diamond \overline{C}} = S$ then $\overline{A \diamond \overline{C}}$ is the largest solution of the equation $A \diamond X = C$.

Proof. Consider a string $\alpha \in (U \cup O)^\star$, then α is in the largest solution of $A \diamond X \subseteq C$ iff $A \diamond \{\alpha\} \subseteq C$ and the following chain of equivalences follows:

$$\begin{aligned}
A \diamond \{\alpha\} \subseteq C &\Leftrightarrow \\
(A_{\Uparrow O} \cap \{\alpha\}_{\Uparrow I})_{\Downarrow I \cup O} \cap \overline{C} = \emptyset &\Leftrightarrow \textit{by Prop. } 2.1(c)\ \overline{C} = (\overline{C}_{\Uparrow U})_{\Downarrow I \cup O} \\
(A_{\Uparrow O} \cap \{\alpha\}_{\Uparrow I})_{\Downarrow I \cup O} \cap (\overline{C}_{\Uparrow U})_{\Downarrow I \cup O} = \emptyset &\Leftrightarrow \textit{by Prop. } 2.5(d) \\
&\quad \textit{since } ((\overline{C}_{\Uparrow U})_{\Downarrow I \cup O})_{\Uparrow U} = \overline{C}_{\Uparrow U} \\
(A_{\Uparrow O} \cap \{\alpha\}_{\Uparrow I} \cap \overline{C}_{\Uparrow U})_{\Downarrow I \cup O} = \emptyset &\Leftrightarrow \textit{by Prop. } 2.7(d) \\
A_{\Uparrow O} \cap \{\alpha\}_{\Uparrow I} \cap \overline{C}_{\Uparrow U} = \emptyset &\Leftrightarrow \textit{by Prop. } 2.7(d) \\
(A_{\Uparrow O} \cap \{\alpha\}_{\Uparrow I} \cap \overline{C}_{\Uparrow U})_{\Downarrow U \cup O} = \emptyset &\Leftrightarrow \textit{by Prop. } 2.5(d) \\
&\quad \textit{since } \{\alpha\}_{\Uparrow I} = ((\{\alpha\}_{\Uparrow I})_{\Downarrow U \cup O})_{\Uparrow I} \\
(\{\alpha\}_{\Uparrow I})_{\Downarrow U \cup O} \cap (A_{\Uparrow O} \cap \overline{C}_{\Uparrow U})_{\Downarrow U \cup O} = \emptyset &\Leftrightarrow \textit{by Prop. } 2.1(c)\ (\{\alpha\}_{\Uparrow I})_{\Downarrow U \cup O} = \{\alpha\} \\
\{\alpha\} \cap (A_{\Uparrow O} \cap \overline{C}_{\Uparrow U})_{\Downarrow U \cup O} = \emptyset &\Leftrightarrow \\
\alpha \notin (A_{\Uparrow O} \cap \overline{C}_{\Uparrow U})_{\Downarrow U \cup O} &\Leftrightarrow \\
\alpha \in \overline{(A_{\Uparrow O} \cap \overline{C}_{\Uparrow U})_{\Downarrow U \cup O}} &\Leftrightarrow \\
\alpha \in \overline{A \diamond \overline{C}}
\end{aligned}$$

Therefore the largest solution of the language equation $A \diamond X \subseteq C$ is given by the language

$$S = \overline{A \diamond \overline{C}}. \tag{2.6}$$

□

Corollary 2.17. *A language B over alphabet $U \cup O$ is a solution of $A \diamond X \subseteq C$ iff $B \subseteq \overline{A \diamond \overline{C}}$.*

Proposition 2.18. *If S is U-convergent, then S is the largest U-convergent solution of the equation, and a language $B \neq \emptyset$ is a U-convergent solution iff $B \subseteq S$.*

When S is not U-convergent the largest U-convergent solution does not exist, since any finite subset of S is a U-convergent solution and therefore no string can be deleted from S without missing a solution. An analogous proposition and remark hold for S-compositionally U-convergent solutions.

2.2.3 *Language Equations Under Bounded Parallel Composition*

Theorem 2.19. *The largest solution of the equation* $A \diamond_l X \subseteq C$ *is the language*

$$S = \overline{(A_{\Uparrow O} \cap \overline{C}_{\Uparrow (U,l)})_{\Downarrow U \cup O}}.$$

Proof.

$$\begin{aligned} A \diamond_l \{\alpha\} \subseteq C &\Leftrightarrow \\ (A_{\Uparrow O} \cap \{\alpha\}_{\Uparrow I} \cap (I \cup O)^{\star}_{\Uparrow (U,l)})_{\Downarrow I \cup O} \cap \overline{C} = \emptyset &\Leftrightarrow \\ A_{\Uparrow O} \cap \{\alpha\}_{\Uparrow I} \cap \overline{C}_{\Uparrow (U,l)} = \emptyset &\Leftrightarrow \\ \alpha \notin (A_{\Uparrow O} \cap \overline{C}_{\Uparrow (U,l)})_{\Downarrow U \cup O} &\Leftrightarrow \\ \alpha \in \overline{(A_{\Uparrow O} \cap \overline{C}_{\Uparrow (U,l)})_{\Downarrow U \cup O}} & \end{aligned}$$

□

2.3 Solution of Equations Over Regular Languages and Finite Automata

Language equations can be solved effectively when they are defined over languages that can be computed with finite procedures. Usually such languages are presented through their corresponding mathematical machines, e.g., finite automata for regular languages. In the following sections, equations over various classes of automata are studied, like FAs and FSMs, specializing the theory of equations to their associated languages. A key issue to investigate is the closure of the solution set with respect to a certain type of language, e.g., when dealing with FSM language equations we require that the solutions are FSM languages. This cannot be taken for granted, because the general solution of abstract language equations is expressed through the operators of complementation and composition, which do not necessarily preserve certain classes of languages.

2.3.1 *An Algorithm to Solve Equations Over Regular Languages and Automata*

Two well-known results [63] are that non-deterministic finite automata are equivalent (with respect to language equality) to deterministic ones and that regular expressions are equivalent to finite automata. By applying the algorithm of subset construction one converts a NDFA into an equivalent DFA (it is complete by construction). Given an NDFA $F = \langle S, \Sigma, \Delta, r, Q \rangle$, the process of subset construction builds the DFA $F' = \langle 2^S, \Sigma, \delta, r, Q' \rangle$, where (1) the states $\tilde{s} \in 2^S$ are the subsets of S, (2) the transition relation is $\delta(i, \tilde{s}) = \cup_{s \in \tilde{s}} \{s' \mid (i, s, s') \in \Delta\}$ and (3) a state is final, i.e., $\tilde{s} \in Q' \subseteq 2^S$, iff $\tilde{s} \cap Q \neq \emptyset$. Since many of the states in 2^S are unreachable from the initial state, they can be deleted and so the determinized automaton usually has fewer states than the power set.[4] To make a NDFA complete it is not necessary to apply the full-blown subset construction, but it suffices to add a new non-accepting state s_d whose incoming transitions are (i, s, s_d) for all i, s for which there was no transition in the original automaton. By a closure construction [63], an NDFA with ϵ-moves can be converted to an NDFA without ϵ-moves; however, a subset construction must be applied at the end to determinize it.

The equivalence of regular expressions and finite automata is shown by matching each operation on regular expressions with a constructive procedure that yields the finite automaton of the result, given the finite automata of the operands. For the most common operations (union, concatenation, complementation, intersection) see [63]. Here we sketch the constructions for the less known operations of projection, lifting, restriction and expansion:

projection ($\downarrow$) Given FA F that accepts language L over $X \times V$, FA F' that accepts language $L_{\downarrow V}$ over X is obtained from F by the following procedure:

replace each edge $((x, v), s, s')$ by the edge (x, s, s').

lifting ($\uparrow$) Given FA F that accepts language L over X, FA F' that accepts language $L_{\uparrow V}$ over $X \times V$ is obtained from F by the following procedure:

replace each edge (x, s, s') by the edges $((x, v), s, s')$, $\forall v \in V$.

restriction ($\Downarrow$) Given FA F that accepts language L over $X \cup V$, FA F' that accepts language $L_{\Downarrow V}$ over V is obtained from F by the following procedure:

$\forall x \in X \setminus V$, change every edge (x, s, s') into the edge (ϵ, s, s'), i.e., replace the symbols $x \in X \setminus V$ by ϵ.[5]

expansion ($\Uparrow$) Given FA F that accepts language L over X, FA F' that accepts language $L_{\Uparrow V}$ over $X \cup V$ $(X \cap V = \emptyset)$ is obtained from F by the following procedure:

for each state s, $\forall v \in V$ add the edge (self-loop) (v, s, s).

l-expansion ($\Uparrow_l$) Given FA F that accepts language L over X, FA F' that accepts language $L_{\Uparrow(V,l)}$, l integer, over $X \cup V$ $(X \cap V = \emptyset)$ is obtained from F by the following procedure:

1. The set of states S' of F' is given by

$$S' = S \cup \{(s, j) \mid s \in S, 1 \leq j \leq l\}.$$

---

[4] It is not uncommon in practice to find that $|\tilde{S}| \leq |S|$.

[5] Apply the closure construction to obtain an equivalent deterministic finite automaton without ϵ-moves.

2. The next state relation Δ' of F' is given by

$$\begin{aligned}\Delta' = \Delta \;\cup\; &\{(v, s, (s, 1)) \mid v \in V, s \in S\}\\ \cup\; &\{(v, (s, j), (s, j+1)) \mid v \in V, s \in S, 1 \leq j < l\}\\ \cup\; &\{(x, (s, j), s') \mid (x, s, s') \in \Delta, 1 \leq j \leq l\}.\end{aligned}$$

3. $r' = r$ and $Q' = Q$.

The procedures for projection, lifting and restriction guarantee the substitution property $f(\epsilon) = \epsilon$.

Given that all the operators used to express the solution of regular language equations have constructive counterparts on automata, we conclude that there is an effective (constructive) way to solve equations over regular languages.

As an example, given a regular language equation $A \bullet X \subseteq C$, where A is a regular language over alphabet $I \times U$, C is over $I \times O$, and the unknown regular language X is over $U \times O$, an algorithm to build X follows.

Procedure 2.3.1. *Input: Regular language equation* $A \bullet X \subseteq C$*; Output: Largest regular language solution* X

1. Consider the finite automata $F(A)$ and $F(C)$ corresponding, respectively, to regular languages A and C.
2. Determinize $F(C)$ by subset construction, if it is a NDFA. The automaton $F(\overline{C})$ of $\overline{C}$ is obtained by interchanging the sets of accepting and non-accepting states of the determinization of $F(C)$.
3. Lift the language A to O by replacing each label (i, u) of a transition of $F(A)$ with all triples (i, u, o), $o \in O$. Lift the language $\overline{C}$ to U by replacing each label (i, o) of a transition of $F(\overline{C})$ with all triples (i, u, o), $u \in U$.
4. Build the automaton $F(A \cap \overline{C})$ of the intersection $A \cap \overline{C}$. The states are pairs of states of the lifted automata $F(A)$ and $F(\overline{C})$, the initial state is the pair of initial states, and a state of the intersection is accepting if both states of the pair are accepting. There is a transition from the state (s_1, s_2) to the state (s_1', s_2') labeled with action (i, u, o) in $F(A \cap \overline{C})$, if there are corresponding transitions labeled with (i, u, o) from state s_1 to state s_1' in $F(A)$ and from s_2 to s_2' in $F(\overline{C})$.
5. Project $F(A \cap \overline{C})$ to $U \times O$ to obtain $F(A \bullet \overline{C})$ by deleting i from the labels (i, u, o). Projection in general makes the finite automaton non-deterministic.
6. Determinize $F(A \bullet \overline{C})$ by subset construction, if it is a NDFA. The automaton $F(\overline{A \bullet \overline{C}})$ corresponding to the regular language solution $X = \overline{A \bullet \overline{C}}$ is obtained by interchanging the sets of accepting and non-accepting states of the determinization of $F(A \bullet \overline{C})$.

Notice that Procedure 2.3.1 holds for any regular language, not only for prefix-closed languages as in restricted versions reported in the literature.

A companion procedure to solve the regular language equation under parallel composition $A \diamond X \subseteq C$ is obtained from Procedure 2.3.1, after replacing the

Cartesian product with union, projection with restriction and lifting with expansion. The largest solution of parallel equations for prefix-closed regular languages had been known already in the process-algebra literature [90, 94, 118].

2.3.2 An Application to Converter Synthesis: The Protocol Mismatch Problem

We apply the algorithm in Sect. 2.3.1 to an equation over finite automata to solve a problem of converter synthesis, i.e., the design of an automaton to translate between two different protocols.

A communication system has a sending part and a receiving part that exchange data through a specific protocol. A mismatch occurs when two systems with different protocols try to communicate. The mismatch problem is solved by designing a converter that translates between the receiver and the sender, while respecting the overall service specification of the behavior of the composed communication system relative to the environment. We formulate the problem as a parallel language equation: given the service specification C of a communication system, a component sender and a component receiver, find a converter X whose composition with the sender and receiver A meets the system specification after hiding the internal signals: $A \diamond X \subseteq C$.

As an example we consider the problem of designing a protocol converter to interface: an *alternating-bit* (AB) sender and a *non-sequenced* (NS) receiver. This problem is adapted from [78] and [56]. A communication system based on an alternating bit protocol is composed of two processes, a sender and a receiver, which communicate over a half duplex channel that can transfer data in either directions, but not simultaneously. Each process uses a control bit called the alternating bit, whose value is updated by each message sent over the channel in either direction. The acknowledgement is also based on the alternating bit: each message received by either process in the system corresponds to an acknowledgement message that depends on the bit value. If the acknowledgement received by a process does not correspond to the message sent originally, the message is resent until the correct acknowledgement is received. On the other hand, a communication system is non-sequenced when no distinction is made among the consecutive messages received or their corresponding acknowledgements. This means that neither messages nor their acknowledgements are distinguished by any flags such as with the alternating bit.

Figure 2.4 shows the block diagram of the composed system. Each component is represented by a rectangle with incoming and outgoing labeled arrows to indicate the inputs and outputs, respectively. The sender consists of an AB protocol sender (*PS*) and of an AB protocol channel (*PC*). Meanwhile, the receiving part includes an NS protocol receiver (*PR*). The converter X must interface the two mismatched protocols and guarantee that its composition with PS, PC and PR refines the service specification (*SS*) of the composed system. The events *Acc* (*Accept*) and *Del*

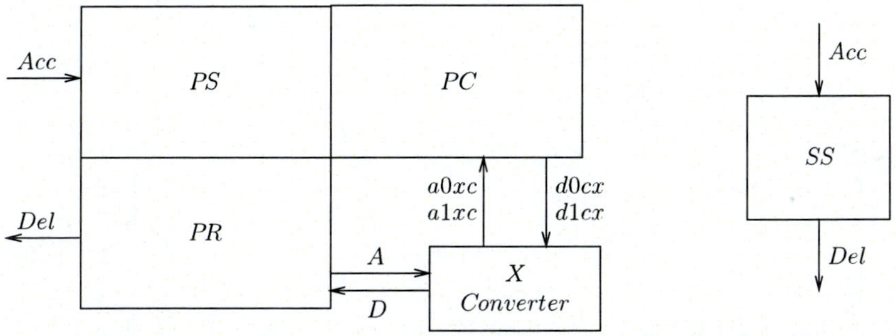

Fig. 2.4 Communication system described in Sect. 2.3.2

(*Deliver*) represent the interfaces of the communication system with the environment (the users). The converter *X* translates the messages delivered by the sender *PS* (using the alternating bit protocol) into a format that the receiver *PR* understands (using the non-sequenced protocol). For example, acknowledgement messages *A* delivered to the converter by the receiver are transformed into acknowledgements of the alternating bit protocol (*a0xc* to acknowledge a 0 bit and *a1xc* to acknowledge a 1 bit) and passed to the sender by the channel (*a0cs* to acknowledge a 0 bit and *a1cs* to acknowledge a 1 bit); data messages are passed from the sender to the channel (*d0sc* for a message controlled by a 0 bit and *d1sc* for a message controlled by a 1 bit) and then from the channel to the converter (*d0cx* for a message controlled by a 0 bit and *d1cx* for a message controlled by a 1 bit) to be transformed by the converter into a data message *D* for the receiver.

We model the components as I/O automata [89], which recognize prefix-closed regular languages, and we solve their language equations. Figure 2.5 shows the automata of the components of the communication system. Missing transitions go to a trap (non-accepting) state, that loops to itself under any event.

Figure 2.6 shows the largest prefix-closed solution $S = \overline{PS \diamond PC \diamond PR \diamond \overline{SS}}$ of the converter problem. All missing transitions go to an *accepting* trap state *dc* (not shown), that would loop to itself under any event; e.g., the initial state would have a transition to state *dc* under events $A, a0xc, a1xc, d1cx$. These transitions are not indicated in the state transition graph of the automaton of the solution language to avoid cluttering the picture. State *dc* can be termed the *don't care* state, because it is introduced during the determinization step to complete the automaton $PS \diamond PC \diamond PR \diamond \overline{SS}$, before the final complementation. It is reached by transitions that cannot occur due to impossible combinations of events in the composition of $PS \diamond PC \diamond PR$ and *S*, and so it does not matter how *S* behaves, once it is in state *dc* (thus the qualification *don't care* state). This makes the largest solution *S* non-deterministic. The solution presented in [78] and [56] does not feature this trap accepting state and so it is not complete (in [78] and [56] all missing transitions of the solution are supposed to end up in a *non-accepting* trap state, a *fail* state); without the above *dc* state, one gets only a subset of all solutions (in particular the complete solutions are missed) and this might lead to an inferior implementation.

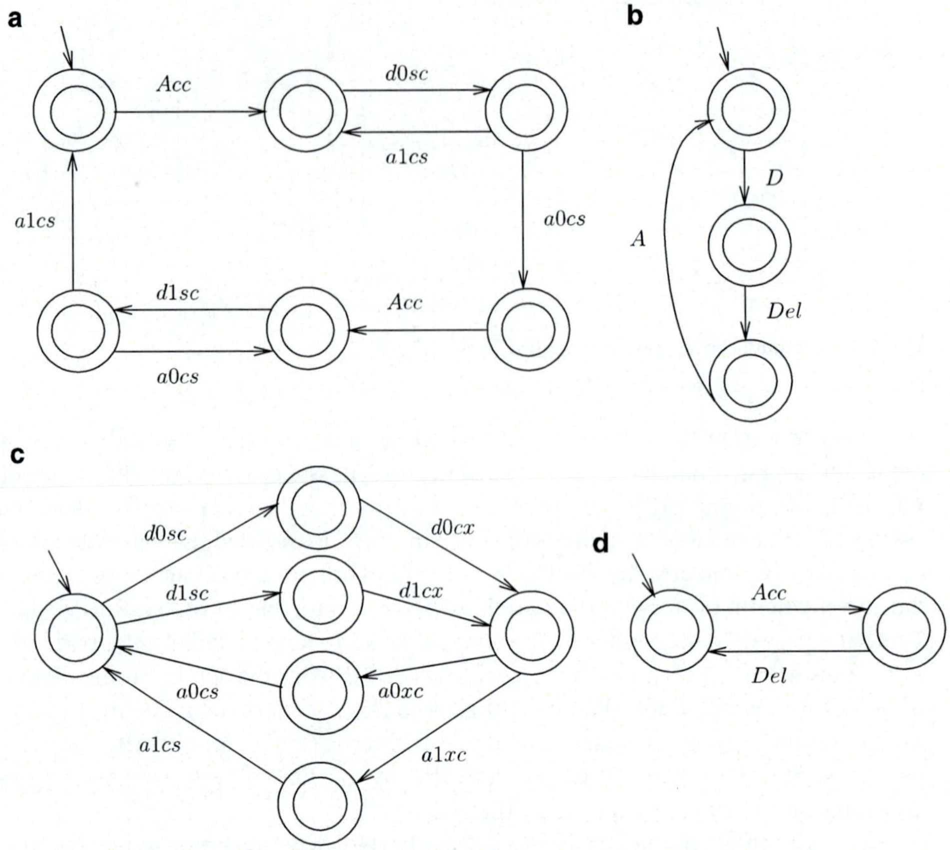

Fig. 2.5 Automata of communication system described in Sect. 2.3.2. (**a**) Automaton of PS; (**b**) automaton of PR; (**c**) automaton of PC; (**d**) automaton of SS

Figure 2.7 shows yet another view of the largest prefix-closed solution $S = \overline{PS \diamond PC \diamond PR \diamond \overline{SS}}$ of the converter problem, with the *dc* state included and the *fail* state excluded.

Figure 2.8 shows the largest prefix-closed 2-bounded solution of the converter problem, as an example of solution with bounded internal communication.

Figure 2.9 shows the composition $PS \diamond PC \diamond PR \cap S_{\Uparrow\{Acc,Del\}}$ of the communication system $PS \diamond PC \diamond PR$ and of the largest converter S. The largest prefix-closed solution S is compositionally $(I^\star O)$-progressive and compositionally prefix $(U \cup V)$-deadlock-free, but not compositionally prefix $(U \cup V)$-convergent. This means that, even if we extract from the largest solution S a complete solution $\hat{S}$ (i.e., $\hat{S}$ is defined for every internal input $A, d0cx, d1cx$), the composition of $\hat{S}$ with the context ($PS \diamond PC \diamond PR$ may deadlock, i.e., the automata can exchange internal actions without producing any output for the environment; this happens, for instance, if $\hat{S}$ after the internal input *d0cx* always selects the internal output *a1cx* (and never selects D). In general the composition of any complete solution that never produces the internal output D with the context has livelocks (loop

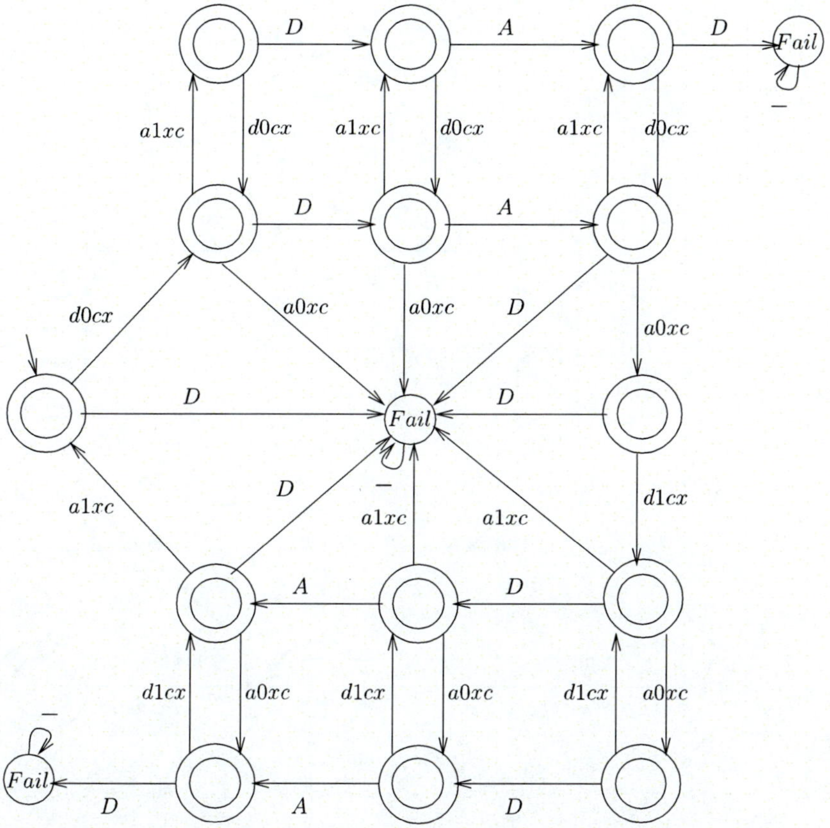

Fig. 2.6 Largest prefix-closed solution $S = \overline{PS \diamond PC \diamond PR \diamond \overline{SS}}$ of the converter problem of Sect. 2.3.2

a1xc-d0cx), which are deadlocks with respect to the external input *Acc*, i.e., the composition is not complete. So the largest solution S is not compositionally prefix $(U \cup V)$-convergent due to the existence of internal cycles, but since the latter can be exited by selecting the internal output D S is compositionally prefix $(U \cup V)$-deadlock-free (and therefore compositionally $(I^\star O)$-progressive).

The protocol conversion problem was addressed in [78], as an instance of supervisory control of discrete event systems, where the converter language is restricted to be a sublanguage of the context A, and in [56] with the formalism of input-output automata. In [78] the problem is modeled by the equation $A \diamond X = C$ over regular languages with the rectification topology (see Fig. 1d). The solution is given as a sublanguage of A of the form $A \diamond C \setminus A \diamond \overline{C}$ (not the largest solution). An algorithm to obtain the largest compositionally progressive solution is provided that first splits the states of the automaton of the unrestricted solution (refining procedure, exponential step due to the restriction operator), and then deletes the states that violate the desired requirement of progressive composition (linear step). This algorithm does not generalize as it is to topologies where the unknown component depends also on signals that do not appear in the component A.

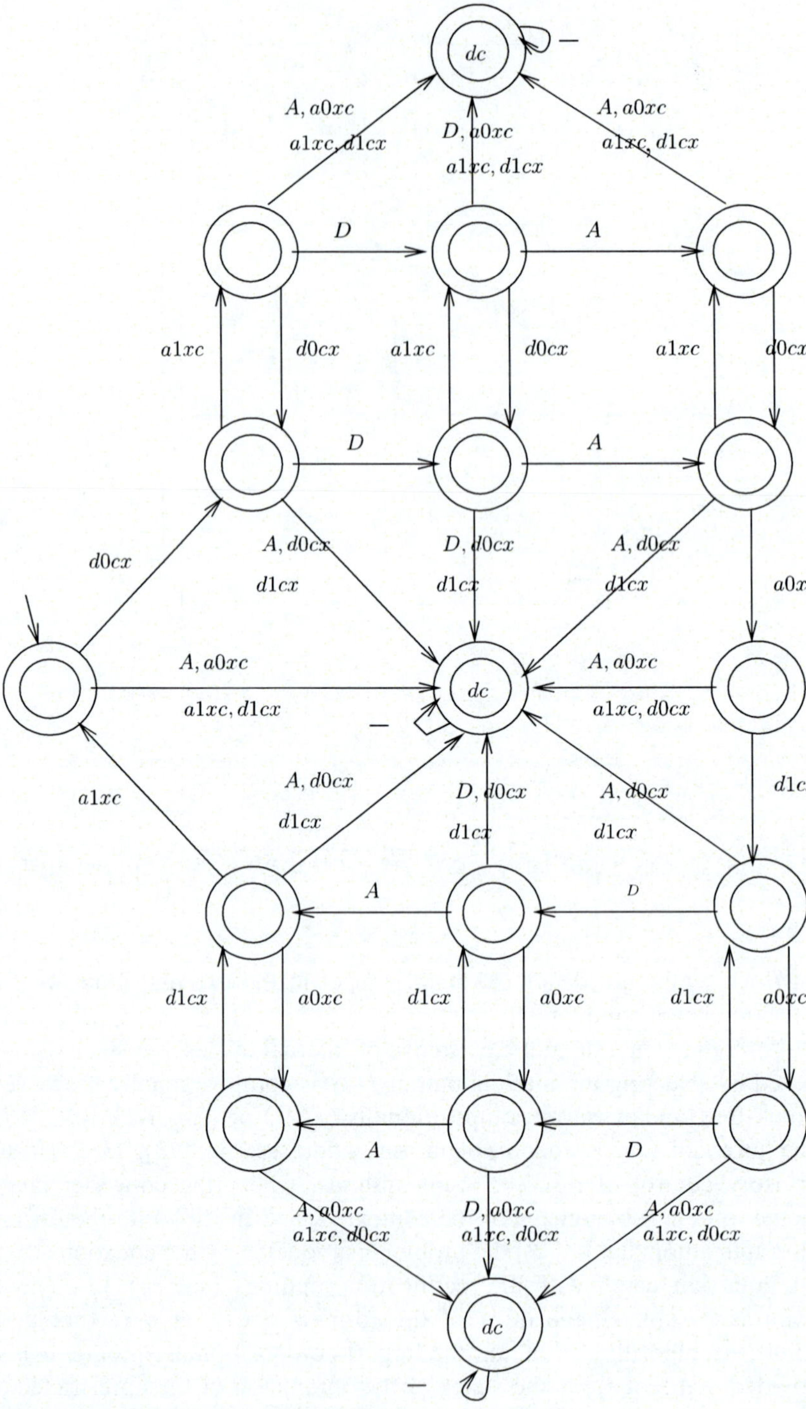

Fig. 2.7 Largest prefix-closed solution S of the converter problem of Sect. 2.3.2. It shows explicitly the transitions to the *dc* state

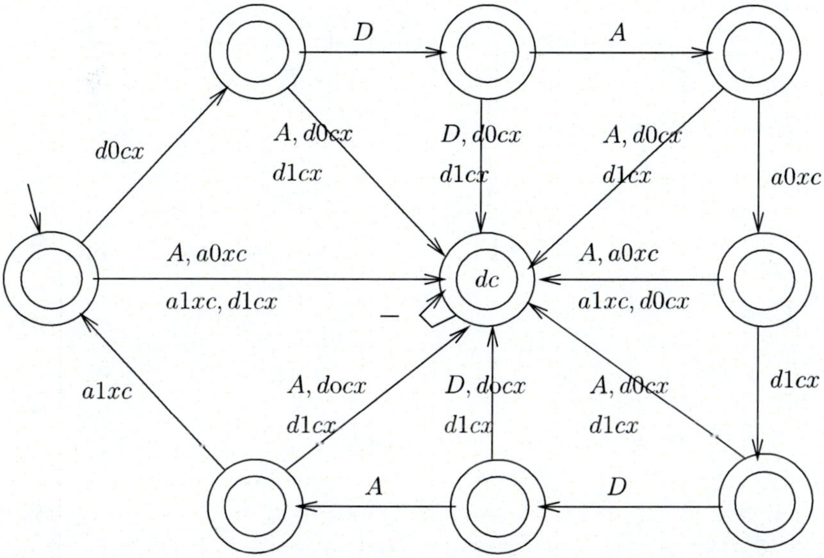

Fig. 2.8 Largest prefix-closed 2-bounded solution of the converter problem of Sect. 2.3.2

Problems

2.1. Consider the traffic controller example described in [12]. For ease of reference we show in Figs. 2.10 and 2.11 respectively the context automaton A and the specification automaton C derived from it.

Notice that the inputs of A are two binary variables v_1, v_2, and the multi-valued variable *colours* that can assume the values *green, red, yellow*. The inputs of C are two multi-valued variables i_1, i_2 that can assume each one of the three values $1, 2, 3$, and again the multi-valued variable *colours*.

Find the largest solution of the equation $A \bullet X \subseteq C$, where the input variables of X are i_1, i_2 and v_1, v_2. This is a series topology where X feeds A, and the composition of X and A yields C.

Repeat the problem assuming that X has one more input: the variable *colours* of automaton A, as in the controller's topology.

Hint.

The largest solution automaton of the series-topology equation is shown in Fig. 2.12.

If the reader will find too cumbersome carrying on by the hand the computations, it will be one more reason to appreciate the automatica tool BALM presented in the second part of the book.

2.2. Consider the alphabets $I = \{i0, i1\}$, $V = \{v0, v1\}$, $O = \{o0, o1\}$, and the automata $A = \langle S_A, \Sigma_A, \Delta_A, r_A, Q_A \rangle$ and $C = \langle S_C, \Sigma_C, \Delta_C, r_C, Q_C \rangle$. Let the automata be defined as follows: $S_A = Q_A = \{sa, sb\}$, $\Sigma_A = V \times O$, $\Delta_A = \{(v0\_o1,\ sa,\ sb), (v1\_o0,\ sa,\ sb), (v0\_o1,\ sb,\ sb), (v1\_o1,\ sb,\ sa)$, $r_A = sa$,

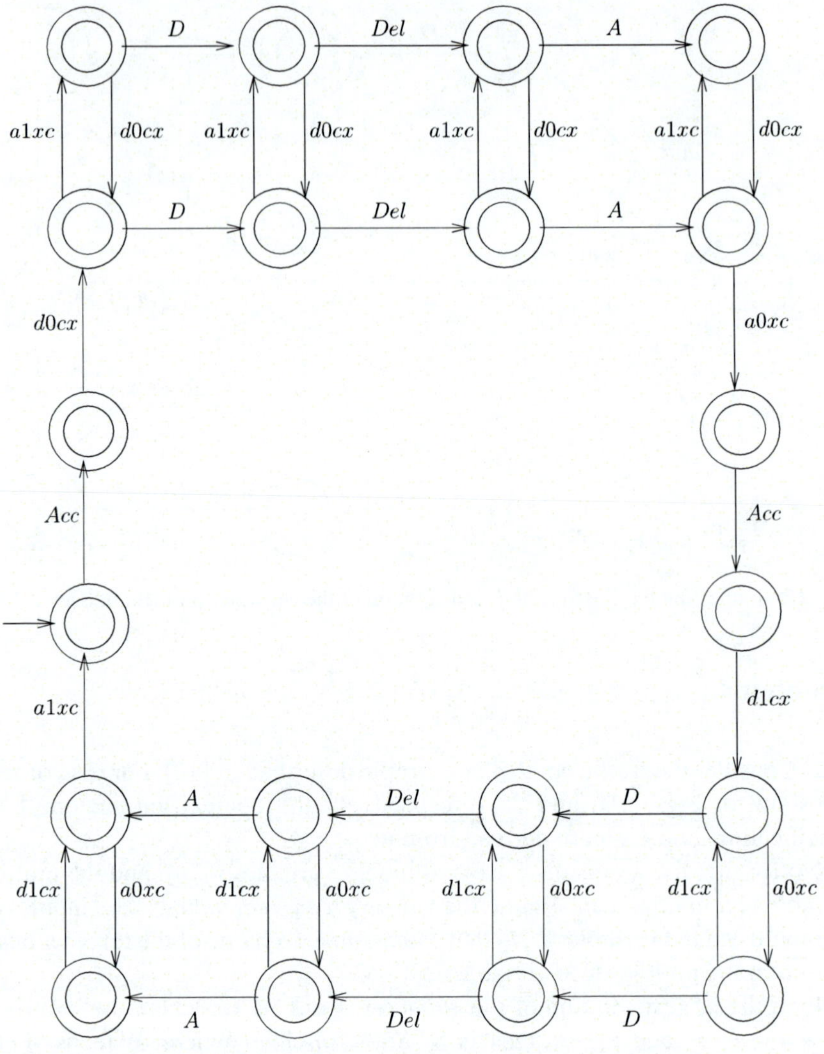

Fig. 2.9 Composition $PS \diamond PC \diamond PR \cap S_{\Uparrow\{Acc,Del\}}$ of communication system $PS \diamond PC \diamond PR$ and largest converter S of the converter problem of Sect. 2.3.2

and $S_C = Q_C = \{s1,\ s2,\ s3,\ s4\}$, $\Sigma_C = I \times O$, $\Delta_C = \{(i1\_o0,\ s1,\ s2)$, $(i0\_o0,\ s1,\ s3)$, $(i0\_o1,\ s2,\ s1)$, $(i1\_01,\ s2,\ s3)\}$, $(i0\_o1,\ s3,\ s1)$, $(i1\_o1,\ s3,\ s4)$, $(i1\_o1,\ s4,\ s3)\}$, $(i0\_o0,\ s4,\ s3)$, $r_C = s1$.

Compute the largest solution of the equation $A \bullet X \subseteq C$, where X has inputs $I \times V$.

Verify whether in this example the largest solution is complete with respect to I.

2.3. Consider the alphabets $I = \{i1, i2\}$, $V = \{v1, v2\}$, $O = \{o1, o2\}$, and the automata $A = \langle S_A, \Sigma_A, \Delta_A, r_A, Q_A\rangle$ and $C = \langle S_C, \Sigma_C, \Delta_C, r_C, Q_C\rangle$. Let the automata be defined as follows: $S_A = Q_A = \{s1, s2\}$, $\Sigma_A = V \times O$, $\Delta_A =$

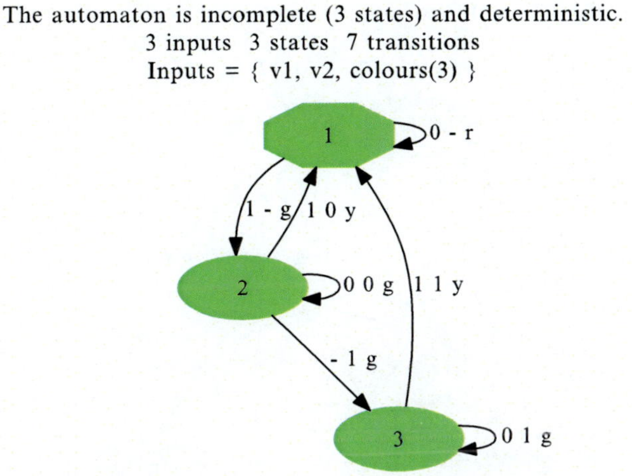

Fig. 2.10 Graphical output of BALM showing the automaton `traffic-ma.aut` describing the fixed automaton component of the traffic light controller problem. The notation of BALM will be presented formally later on, but this graph can be read as any usual FSM graphical representation

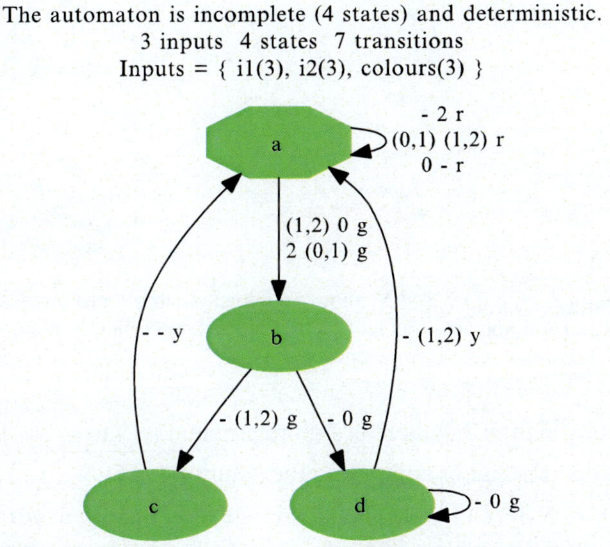

Fig. 2.11 Graphical output of BALM showing the automaton `traffic-mc.aut` describing the specification automaton of the traffic light controller problem

$\{(v1\_o1,\ s1,\ s2), (v2\_o2,\ s1,\ s1), (v1\_o2,\ s2,\ s1), (v2\_o1,\ s2,\ s2)$, $r_A = s1$, and $S_C = Q_C = \{sa,\ sb,\ sc\}$, $\Sigma_C = I \times O$, $\Delta_C = \{(i1\_o1,\ sa,\ sb), (i2\_o2,\ sa,\ sc), (i2\_o2,\ sb,\ sa), (i1\_02,\ sb,\ sc)\}, (i1\_o1,\ sc,\ sa), (i2\_o1,\ sc,\ sb)$, $r_C = sa$.

Compute the largest solution of the equation $A \bullet X \subseteq C$, where X has inputs $I \times V \times O$.

Verify whether in this example of controller's topology the composition of the largest solution with M_A is progressive.

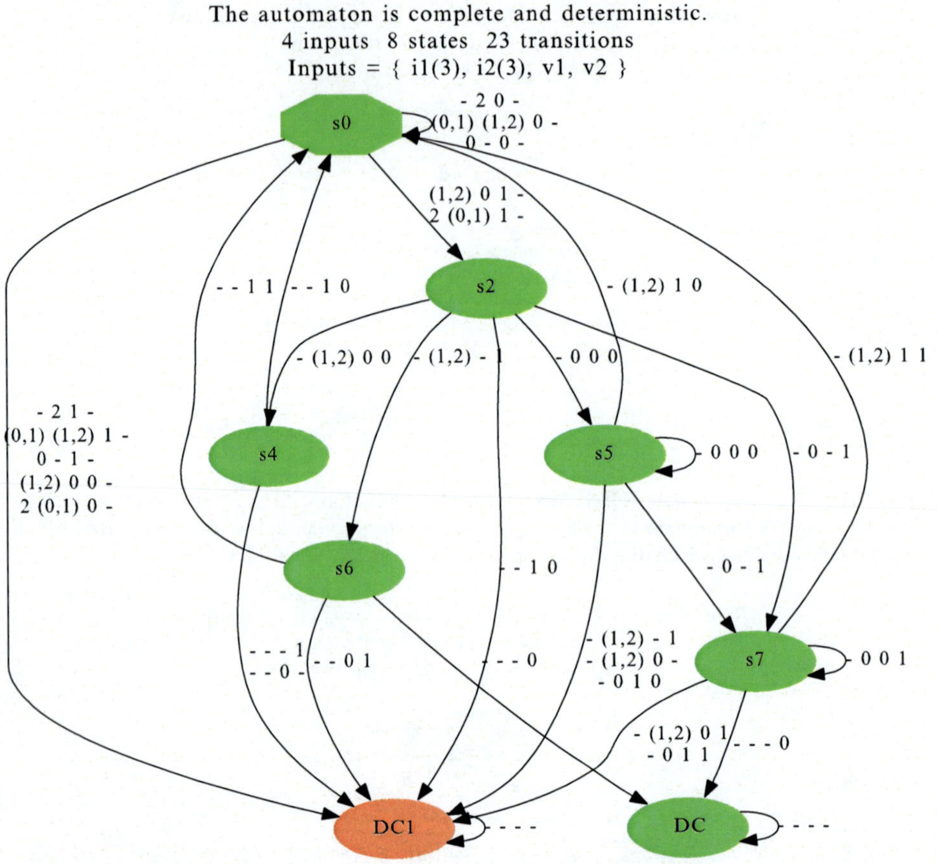

Fig. 2.12 Graphical output of BALM showing the automaton `traffic-xfsm_min.aut` describing the largest solution automaton of the traffic light controller problem

2.4. Somebody claims that there is a more efficient way to solve the language equation $F \bullet X \subseteq S$. Instead of computing the solution $\overline{(F \cap \overline{S}_{\uparrow U \times V})_{\downarrow U \times V}}$, compute as a solution $(\overline{F \cap \overline{S}_{\uparrow U \times V}})_{\downarrow U \times V}$. The advantage would be to avoid a determinization because the second complementation is performed before the projection (the latter introduces non-determinism and so the need for determinization). Is it correct? Are the two expressions equivalent ? Is one contained in the other?

Chapter 3
Equations Over Finite State Machines

3.1 Finite State Machines and Their Languages

3.1.1 Finite State Machines

Definition 3.1.1. A **finite state machine** (FSM) is a 5-tuple $M = \langle S, I, O, T, r \rangle$ where S represents the finite state space, I represents the finite input space, O represents the finite output space and $T \subseteq I \times S \times S \times O$ is the transition relation. On input i, the FSM at present state p may transit to next state n and produce output o iff $(i, p, n, o) \in T$. State $r \in S$ represents the initial or reset state. We denote the projection of relation T to $I \times S \times S$ (next state relation) by $T_n \subseteq I \times S \times S$, i.e., $(i, s, s') \in T_n \Leftrightarrow \exists o \; (i, s, s', o) \in T$; similarly, we denote the projection of relation T to $I \times S \times O$ (output relation) by $T_o \subseteq I \times S \times O$, i.e., $(i, s, o) \in T_o \Leftrightarrow \exists s' \; (i, s, s', o) \in T$. Sometimes δ is used instead of T_n and λ instead of T_o.

If at least one transition is specified for each present state and input pair, the FSM is said to be **complete**. If no transition is specified for at least one present state and input pair, the FSM is said to be **partial**.

An FSM is said to be **trivial** when $T = \emptyset$, denoted by M_ϵ.

The notion of a partial FSM should not be confused with that of an incompletely specified FSM (ISFSM), which is traditionally employed in the digital design literature to denote a form of restricted non-determinism widely used in the specification of sequential logic designs (see [66]). ISFSMs are introduced in Definition 3.1.4.

It is convenient to think of the relations T_n and T_o as functions $T_n : I \times S \to 2^S$ and $T_o : I \times S \to 2^O$.

Definition 3.1.2. An FSM $M' = \langle S', I', O', T', r' \rangle$ is a **submachine** of FSM $M = \langle S, I, O, T, r \rangle$ if $S' \subseteq S$, $I' \subseteq I$, $O' \subseteq O$, $r' = r$, and T' is a restriction of T to the domain of definition $I' \times S' \times S' \times O'$.

T. Villa et al., *The Unknown Component Problem: Theory and Applications*,
DOI 10.1007/978-0-387-68759-9_3,

Definition 3.1.3. A **deterministic FSM** (DFSM) is an FSM where for each pair $(i, p) \in I \times S$, there is at most one next state n and one output o such that $(i, p, n, o) \in T$, i.e., there is at most one transition from p under i. An FSM that is not a DFSM is a **non-deterministic finite state machine** (NDFSM).

In a DFSM the next state n and the output o can be given, respectively, by a next state function $n = T_n(i, p)$ and an output function $o = T_o(i, p)$, where T_n and T_o may be partial functions when the DFSM is partial.

Definition 3.1.4. An **incompletely specified FSM** (ISFSM) is a NDFSM such that, for a given input and present state, either there is a unique next state and output, or the next state is a designated don't care state DNC and any output is produced; moreover, at the state DNC under any input there is a self-loop and any output is produced.

Definition 3.1.5. An NDFSM is a **pseudo non-deterministic FSM** (PNDFSM) [142], or **observably non-deterministic FSM**[28], or **observable FSM**[130], if for each triple $(i, p, o) \in I \times S \times O$, there is at most one state n such that $(i, p, n, o) \in T$.

The qualification "non-deterministic" is because for a given input and present state, there may be more than one possible output; however, edges (i.e., transitions) carrying different outputs must go to different next states. The further qualification "pseudo" non-deterministic is because its underlying finite automaton is deterministic. In a PNDFSM the next state n, if it exists, is unique for a given combination of input, present state and output, so it can be defined by a partial next state function $n = T_n(i, p, o)$. The output is represented by a relation $T_o \subseteq I \times S \times O$, because the output is non-deterministic in general.

Definition 3.1.6. A complete FSM is said to be of **Moore** type if $(i, p, n, o) \in T$ implies that for all i' there is n' such that $(i', p, n', o) \in T$.[1]

The transition relation T of an FSM can be extended in the usual way to a relation on $I^\star \times S \times S \times O^\star$: given a present state p and an input sequence $i_1 \dots i_k \in I^\star$, $(i_1 \dots i_k, p, n, o_1 \dots o_k) \in T$ iff there is a sequence $s_1 \dots s_{k+1}$ such that $s_1 = p, \dots, s_{k+1} = n$ and for each $j = 1, \dots, k$ it holds that $(i_j, s_j, s_{j+1}, o_j) \in T$. A similar extension can be defined for T_p and T_n.

Here FSMs are assumed to be pseudo non-deterministic, unless otherwise stated. It is always possible to convert a general NDFSM into a PNDFSM by subset construction.

[1]Notice that this definition allows for NDFSMs of Moore type, contrary to the more common definition of Moore type: for each present state p there is an output o such that all transitions whose present state is p carry the same output o.

3.1.2 Languages of FSMs

We now introduce the notion of a language associated to an FSM. This is achieved by looking to the automaton underlying a given FSM. For our purposes, we define two related languages: one over the alphabet $I \times O$ and the other over the alphabet $I \cup O$, naturally associated, respectively, with synchronous and parallel composition, as it will be seen later.

For a language over $I \times O$, the automaton coincides with the original FSM where all states are made accepting and the edges carry a label of the type (i, o).

For a language over $I \cup O$, the automaton is obtained from the original FSM, by replacing each edge (i, s, s', o) by the pair of edges $(i, s, (s, i))$ and $(o, (s, i), s')$ where (s, i) is a new node (non-accepting state). All original states are made accepting. The automaton is deterministic because from (i, s, s'_1, o_1) and (i, s, s'_2, o_2) the edges $(i, s, (s, i))$, $(o_1, (s, i), s'_1)$ and $(o_2, (s, i), s'_2)$ are obtained (the same edge $(i, s, (s, i))$ works in both cases).

Definition 3.1.7. Given an FSM $M = \langle S, I, O, T, r\rangle$, consider the finite automaton $F(M) = \langle S, I \times O, \Delta, r, S\rangle$, where $((i, o), s, s') \in \Delta$ iff $(i, s, s', o) \in T$. The language accepted by $F(M)$ is denoted $L_r^\times(M)$, and by definition is the $\times$**-language** of M at state r. Similarly $L_s^\times(M)$ denotes the language accepted by $F(M)$ when started at state s, and by definition is the $\times$-language of M at state s.

Definition 3.1.8. Given an FSM $M = \langle S, I, O, T, r\rangle$, consider the finite automaton $F(M) = \langle S \cup (S \times I), I \cup O, \Delta, r, S\rangle$, where $(i, s, (s, i)) \in \Delta \wedge (o, (s, i), s') \in \Delta$ iff $(i, s, s', o) \in T$. The language accepted by $F(M)$ is denoted $L_r^\cup(M)$, and by definition is the $\cup$**-language** of M at state r. Similarly $L_s^\cup(M)$ denotes the language accepted by $F(M)$ when started at state s, and by definition is the $\cup$-language of M at state s. By construction, $L_s^\cup(M) \subseteq (IO)^\star$, where IO denotes the set $\{io \mid i \in I, o \in O\}$.

In both cases, $\epsilon \in L_r(M)$ because the initial state is accepting. An FSM M is **trivial** iff $L_r(M) = \{\epsilon\}$.

Definition 3.1.9. A language L is an **FSM language** if there is an FSM M such that the associated automaton $F(M)$ accepts L. The language associated to a DFSM is sometimes called a **behavior**.[2]

Remark. When convenient, we will say that FSM M has property X if its associated FSM language has property X.

Definition 3.1.10. State t of FSM M_B is said to be a **reduction** of state s of FSM M_A (M_A and M_B are assumed to have the same input/output set), written $t \preceq s$,

[2] The language associated to a NDFSM includes a set of behaviors.

iff $L_t(M_B) \subseteq L_s(M_A)$. States t and s are **equivalent states**, written $t \cong s$, iff $t \preceq s$ and $s \preceq t$, i.e., when $L_t(M_B) = L_s(M_A)$. An FSM with no two equivalent states is a **reduced** FSM.

Similarly, M_B is a **reduction** of M_A, $M_B \preceq M_A$, iff r_{M_B}, the initial state of M_B, is a reduction of r_{M_A}, the initial state of M_A. When $M_B \preceq M_A$ and $M_A \preceq M_B$ then M_A and M_B are **equivalent FSMs**, i.e., $M_A \cong M_B$.

For complete DFSMs reduction and equivalence of states coincide. An FSM can be reduced by merging the equivalent states with the classical state minimization procedure [63]. Given an FSM language, there is a family of equivalent FSMs associated with it; for simplicity we will usually speak of the FSM associated with a given FSM language. In this paper, complete deterministic FSMs are assumed to be reduced, unless stated otherwise.

An FSM language is regular, whereas the converse is not true.

Theorem 3.1. *A regular language over alphabet $I \times O$ is the language of a complete FSM over input alphabet I and output alphabet O iff L is prefix-closed and I-progressive. A regular language that is prefix-closed, but not I-progressive, is the language of a partial FSM.*

Notice that the merging of the notions of complete FSM and I-progressive associated language is due to the fact that FSMs are assumed to be PNDFSMs, i.e., their underlying automaton is deterministic, therefore a word has a unique run (sequence of transitions), from which an extension is possible under any input.

Theorem 3.2. *A regular language over alphabet $I \cup O$ is the language of a complete FSM over input alphabet I and output alphabet O iff $L \subseteq (IO)^\star$, L is IO-prefix-closed and IO-progressive. A regular language $L \subseteq (IO)^\star$ that is IO-prefix-closed, but not IO-progressive, is the language of a partial FSM.*

Given a regular language L over alphabet $I \times O$, an algorithm follows to build L^{FSM}, the largest subset of L that is the $\times$-language of an FSM over input alphabet I and output alphabet O.

Procedure 3.1.1. *Input: Regular language L over $I \times O$; Output: Largest FSM language $L^{FSM} \subseteq L$.*

1. Build a deterministic automaton A accepting L.
2. Delete all nonfinal states together with their incoming edges.
3. If the initial state has been deleted, then $L^{FSM} = \emptyset$. Otherwise, let $\hat{A}$ be the automaton produced by the procedure and L^{FSM} the language that $\hat{A}$ accepts. If there is no outgoing edge from the initial state of $\hat{A}$, then $\hat{A}$ accepts the trivial FSM language $L^{FSM} = \{\epsilon\}$, otherwise it accepts a nontrivial FSM language L^{FSM}. Any FSM language in L must be a subset of L^{FSM}.

In general this procedure leads to a partial FSM. To obtain the largest subset of L that is the language of a complete FSM we must apply one more pruning algorithm.

Procedure 3.1.2. *Input: FSM Language L^{FSM} over $I \times O$; Output: Largest I-progressive FSM language $Prog(L^{FSM}) \subseteq L^{FSM}$.*

1. Build a deterministic automaton A accepting L^{FSM}.
2. Iteratively delete all states that have an undefined transition for some input (meaning: states such that $\exists i \in I$ with no $o \in O$ for which there is an outgoing edge carrying the label (i, o)), together with their incoming edges, until the initial state is deleted or no more state can be deleted.
3. If the initial state has been deleted, then $Prog(L^{FSM}) = \emptyset$. Otherwise, let $\hat{A}$ be the automaton produced by the procedure and $Prog(L^{FSM})$ the language that $\hat{A}$ accepts. Any I-progressive FSM language in L^{FSM} must be a subset of $Prog(L^{FSM})$.

Theorem 3.3. *Procedure 3.1.2 returns the largest I-progressive subset of L^{FSM}.*

Proof. Define a state s of the automaton A representing L^{FSM} as I_1-nonprogressive if for some $i \in I$ and for all $o \in O$ there is no state reached from s under a transition labeled with (i, o). State s is I_k-nonprogressive, $k > 1$, if for some $i \in I$ and for all $o \in O$ each state reached from s under the transition labeled (i, o) is I_j-nonprogressive, $j < k$. State s is I-nonprogressive if it is I_k-nonprogressive for some $k \geq 1$. The language $Prog(L^{FSM})$ is represented by the automaton $Prog(A)$, obtained from A by removing iteratively the I-nonprogressive states and the related transitions.

We must prove that if $K \subseteq L^{FSM}$ and K is I-progressive then $K \subseteq Prog(L^{FSM})$. The proof goes by induction. If K is I-progressive, there is no string in K that takes the automaton A from the initial state to an I_1-nonprogressive state. Suppose now by induction hypothesis that no string in K takes A to an I_k-nonprogressive state, $k \geq 1$. We must conclude that, if K is I-progressive, there is also no string that takes A to an I_{k+1}-nonprogressive state, otherwise, by definition of I_{k+1}-nonprogressive, K has a string that takes A to some I_j-nonprogressive state, $j \leq k$. Therefore no string in K takes the automaton A to a nonprogressive state, i.e., $K \subseteq Prog(L^{FSM})$. □

Proposition 3.4. *An FSM whose language is L^{FSM} or $Prog(L^{FSM})$ can be deduced trivially from $\hat{A}$ (obtained according to Procedure 3.1.2) by interpreting each label (i, o) as an input/output pair i/o.*

Proposition 3.5. *Given a regular language L over alphabet $I \times O$, let M be an FSM over input alphabet I and output alphabet O. The language $L_r^{\times}(M)$ of M is contained in L iff $L_r^{\times}(M) \subseteq L^{FSM}$.*

Proof. Show that $L_r^{\times}(M) \subseteq L \Rightarrow L_r^{\times}(M) \subseteq L^{FSM}$. Indeed $L_r^{\times}(M)$ is an FSM language contained in L and L^{FSM} is by construction the largest FSM language contained in L. So $L_r^{\times}(M) \subseteq L^{FSM}$.

$L_r^{\times}(M) \subseteq L^{FSM} \Rightarrow L_r^{\times}(M) \subseteq L$, since by definition $L^{FSM} \subseteq L$. □

Given a regular language L over alphabet $I \cup O$, an algorithm follows to build L^{FSM}, the largest subset of L that is the $\cup$-language of an FSM over input alphabet I and output alphabet O.

Procedure 3.1.3. *Input: Regular language L over $I \cup O$; Output: Largest FSM language $L^{FSM} \subseteq L$.*

1. Build a deterministic automaton A accepting $L \cap (IO)^{\star}$.
2. Delete the initial state if it is a nonfinal state.
3. Delete all nonfinal states having incoming edges labeled with symbols from alphabet O, together with their incoming edges.
4. If the initial state has been deleted, then $L^{FSM} = \emptyset$. Otherwise, let $\hat{A}$ be the automaton produced by the procedure and L^{FSM} the language that $\hat{A}$ accepts. If there is no outgoing edge from the initial state of $\hat{A}$, then $\hat{A}$ accepts the trivial language $L^{FSM} = \{\epsilon\}$, otherwise it accepts a nontrivial FSM language L^{FSM}. Any FSM language in L must be a subset of L^{FSM}.

To obtain the largest subset of L that is the language of a complete FSM (IO-progressive) we must apply one more pruning algorithm.

Procedure 3.1.4. *Input: FSM Language L^{FSM} over $I \cup O$; Output: Largest IO-progressive FSM language $Prog(L^{FSM}) \subseteq L^{FSM}$.*

1. Build a deterministic automaton A accepting L^{FSM}.
2. Iteratively delete all states that are final and for which $\exists i \in I$ with no outgoing edge carrying the label i, together with their incoming edges, until the initial state is deleted or no more state can be deleted. Delete the initial state if $\exists i \in I$ with no outgoing edge carrying the label i.
3. If the initial state has been deleted, then $Prog(L^{FSM}) = \emptyset$. Otherwise, let $\hat{A}$ be the automaton produced by the procedure and $Prog(L^{FSM})$ the language that $\hat{A}$ accepts. Any IO-progressive FSM language in L^{FSM} must be a subset of $Prog(L^{FSM})$.

Theorem 3.6. *Procedure 3.1.4 returns the largest IO-progressive subset of L^{FSM}.*

Proof. Similar to the proof of Theorem 3.3. □

Proposition 3.7. *An FSM whose language is L^{FSM} or $Prog(L^{FSM})$ can be deduced trivially from $\hat{A}$ (obtained according to Procedure 3.1.4) by replacing pairs of consecutive edges labeled, respectively, with i and o by a unique edge labeled i/o.*

Proposition 3.8. *Given a regular language L over alphabet $I \cup O$, let M be an FSM over input alphabet I and output alphabet O. The language $L_r^{\cup}(M)$ of M is contained in L iff $L_r^{\cup}(M) \subseteq L^{FSM}$.*

The proof is the same as the one of Prop. 3.5.

Finally we characterize the Moore FSMs that are the reduction of a given FSM. Notice that the language of a Moore FSM is a Moore language.

Procedure 3.1.5. *Input: Complete FSM M; Output: Largest submachine of M that is a Moore FSM, denoted by $Moore(M)$, if it exists.*

Given a state $s \in M$, define the set $K_s = \{o \in O \mid \forall i \in I\ \exists s' \in M\ s.t.\ (i, s, s', o) \in T_M\}$, $K_s \subseteq O$.

1. Iterate for each state $s \in M$ until M does not change.
   (a) Compute the set $K_s \subseteq O$.
   (b) If $K_s \neq \emptyset$ delete from T_M each transition (i, s, s', o) such that $o \notin K_s$; if $K_s = \emptyset$ delete s with all its incoming edges from M.
2. If the initial state has been deleted then there is no submachine of M that is a Moore FSM, otherwise $Moore(M) = M$.

Theorem 3.9. *Any Moore FSM M' that is a reduction of M is a reduction of $Moore(M)$, the output of Procedure 3.1.5.*

Proof. Define a state s of FSM M as 1-nonMoore if $K_s = \emptyset$. State s is k-nonMoore, $k > 1$, if for some $i \in I$ and for all $o \in O$ each state reached from s under the transition labeled (i/o) is j-nonMoore, $j < k$. State s is nonMoore if it is k-nonMoore for some $k \geq 1$. $Moore(M)$ is obtained from M by removing iteratively the nonMoore states and, from the remaining states s, the transitions (i, s, s', o) such that $o \notin K_s$. Notice that by construction $Moore(M)$ is guaranteed to be complete.

We must prove that if $L(M') \subseteq L(M)$ and $L(M')$ is Moore then $L(M') \subseteq L(Moore(M))$. The proof goes by induction. If $L(M')$ is Moore, there is no string in $L(M')$ that takes the FSM M from the initial state to a 1-nonMoore state. Suppose now by induction hypothesis that no string in $L(M')$ takes M to a k-nonMoore state, $k > 1$. We must conclude that, if $L(M')$ is Moore, there is also no string that takes M to a $(k + 1)$-nonMoore state, otherwise, by definition of $(k + 1)$-nonMoore, $L(M')$ has a string that takes M to some j-nonMoore state, $j \leq k$. Therefore no string in $L(M')$ takes the FSM M to a nonMoore state, i.e., $L(M') \subseteq L(Moore(M))$. □

Moore machines play a role in guaranteeing that the composition of FSMs is a complete FSM (see Theorem 3.12).

3.2 Composition of Finite State Machines

Different types of composition between pairs of FSMs may be defined, according to the protocol by which signals are exchanged. For a given composition operator and pair of FSMs we must establish whether the composition of this pair is defined, meaning that it yields a set of behaviors that can be described by another FSM. In general, the composition of FSMs is a partially specified function from pairs of FSMs to an FSM. In our approach we define the composition of FSMs by means of the composition operators over languages introduced in Sect. 2.1. Thus the FSM yielded by the composition of FSMs M_A and M_B is the one whose language is obtained by the composition of the FSM languages associated with M_A and M_B. The synchronous composition operator models the synchronous connection of sequential circuits, while the parallel composition operator models an exchange protocol by which an input is followed by an output after a finite exchange of internal signals. The latter model, introduced in [107], abstracts a system with two components and a single message in transit. At any moment either the components exchange messages or one of them communicates with its environment. The environment submits the next external input to the system only after the system has produced an external output in response to the previous input.

3.2.1 Synchronous Composition of FSMs

Consider the pair of FSMs:

1. FSM M_A has input alphabet $I_1 \times V$, output alphabet $U \times O_1$, and transition relation T_A.
2. FSM M_B has input alphabet $I_2 \times U$, output alphabet $V \times O_2$, and transition relation T_B.

We define a synchronous composition operator $\bullet$ that associates with a pair of FSMs M_A and M_B another FSM $M_A \bullet M_B$ such that:

1. The external input alphabet is $I_1 \times I_2 = I$.
2. The external output alphabet is $O_1 \times O_2 = O$.

The topology for this is shown in Fig. 1a (a variant is in Fig. 1f).[3] Recall that, by definition of synchronous composition of languages, a sequence $\alpha \in (I_1 \times I_2 \times O_1 \times O_2)^\star$ is in the language of the synchronous composition of $L(M_A)$ and $L(M_B)$ iff α is in the projection onto $I_1 \times I_2 \times O_1 \times O_2$ of the intersection of the lifting of $L(M_A)$ over $I_2 \times O_2$ and of the lifting of $L(M_B)$ over $I_1 \times O_1$[4]:

$$\alpha \in L(M_A) \bullet L(M_B) \text{ iff}$$

$$\alpha \in [L(M_A)_{\uparrow I_2 \times O_2} \cap L(M_B)_{\uparrow I_1 \times O_1}]_{\downarrow I \times O}.$$

Notice that the liftings $L(M_A)_{\uparrow I_2 \times O_2}$ and $L(M_B)_{\uparrow I_1 \times O_1}$ are needed to have the languages of M_A and M_B defined on the same alphabet; e.g., $L(M_B)$ is defined over $I_2 \times U \times V \times O_2$, and the lifting $\uparrow I_1 \times O_1$ defines it over $I_1 \times I_2 \times U \times V \times O_1 \times O_2$.

[3] Notice that more complex topologies can be handled in the same way, by defining the compositions with respect to the appropriate sets of variables, as pointed out when introducing the composition of automata.

[4] Use the same order $I_1 \times I_2 \times U \times V \times O_1 \times O_2$ in the languages $L(M_A)_{\uparrow I_2 \times O_2}$ and $L(M_B)_{\uparrow I_1 \times O_1}$.

In this way we describe the behavior of each FSM as a component of the synchronous composition.

Lemma 3.10. *If $L(M_A)$ and $L(M_B)$ are FSM $\times$-languages, then $L(M_A) \bullet L(M_B)$ is an FSM $\times$-language.*

Proof. $L(M_A) \bullet L(M_B)$ is prefix-closed, because prefix-closed FSM $\times$-languages are closed under $\bullet$ composition. Notice that $L(M_A) \bullet L(M_B)$ does not need to be progressive, because partial FSMs are allowed. □

Therefore we can state the following definition.

Definition 3.2.1. The **synchronous composition of FSMs** M_A and M_B yields the FSM $M_A \bullet M_B$ with language

$$L(M_A \bullet M_B) = L(M_A) \bullet L(M_B).$$

If the language $L(M_A) \bullet L(M_B) = \{\epsilon\}$, then $M_A \bullet M_B$ is a trivial FSM.

The previous definition is sound because the language $L(M_A) \bullet L(M_B)$ by Lemma 3.10 is an FSM language, which corresponds to a (partial) complete FSM if the language $L(M_A) \bullet L(M_B)$ is (not) I-progressive. Then by subset construction and state minimization, we produce a reduced observable FSM. In summary, we convert from the FSMs M_A and M_B to the automata accepting their FSM languages, operate on them and then convert back from the resulting automaton to an FSM; then we produce a reduced PNDFSM (we assume that M_A and M_B are PNDFSMs), because subset construction determinizes the underlying finite automaton.

Example 3.11. (a) Synchronous composition of two FSMs defining a complete FSM.

Given the topology shown in Fig. 1.1d, consider the FSMs $M_A = \langle S_A, I_1 \times V, U \times O_1, T_A, s1\rangle$ and $M_B = \langle S_B, U, V, T_B, sa\rangle$ with $S_A = \{s1, s2, s3\}$, $T_A = \{(1\ \text{-},\ s1,\ s1,\ 11), (00.\ s1,\ s2,\ 10), (01,\ s1,\ s3,\ 10), (\text{-}\ 0,\ s2,\ s1,\ 01), (\text{-}\ 1,\ s2,\ s3,\ 10), (\text{-}\ 1,\ s3,\ s1,\ 01), (\text{-}\ 0,\ s3,\ s2,\ 00)\}$, $S_B = \{sa, sb\}$, $T_B = \{(0,\ sa,\ sa,\ 1), (1,\ sa,\ sb,\ 0), (0,\ sb,\ sa,\ 0), (1,\ sb,\ sb,\ 0)\}$.[5]

Then $M_A \bullet M_B = M_{A \bullet B} = \langle S_{A \bullet B}, I_1, O_1, T_{A \bullet B}, (s1, sa)\rangle$ with $S_{A \bullet B} = \{(s1, sa), (s1, sb), (s2, sb)\}$ and $T_{A \bullet B} = \{(1,\ (s1, sa),\ (s1, sb),\ 1), (0,\ (s1, sa),\ (s2, sb),\ 0), (1,\ (s1, sb),\ (s1, sb),\ 1)\ (0,\ (s1, sb),\ (s2, sb),\ 0), (\text{-},\ (s2, sb),\ (s1, sa),\ 1)\}$ is a complete FSM.

Figure 3.1 shows FSMs M_A, M_B and $M_A \bullet M_B$.

(b) Synchronous composition of two FSMs defining a partial FSM.

Modify the transition relation of M_B to obtain M'_B as follows: $T'_B = \{(0,\ sa,\ sa,\ 1), (1\ sa\ sb\ 0), (0,\ sb,\ sa,\ 1), (1,\ sb,\ sb,\ 0)\}$. Then $T_{A \bullet B'} = \{(1,\ (s1, sa),\ (s1, sb),\ 1), (0,\ (s1, sa),\ (s2, sb),\ 0), (1,\ (s1, sb),\ (s1, sb),\ 1)$

[5]— denotes input or output don't care conditions.

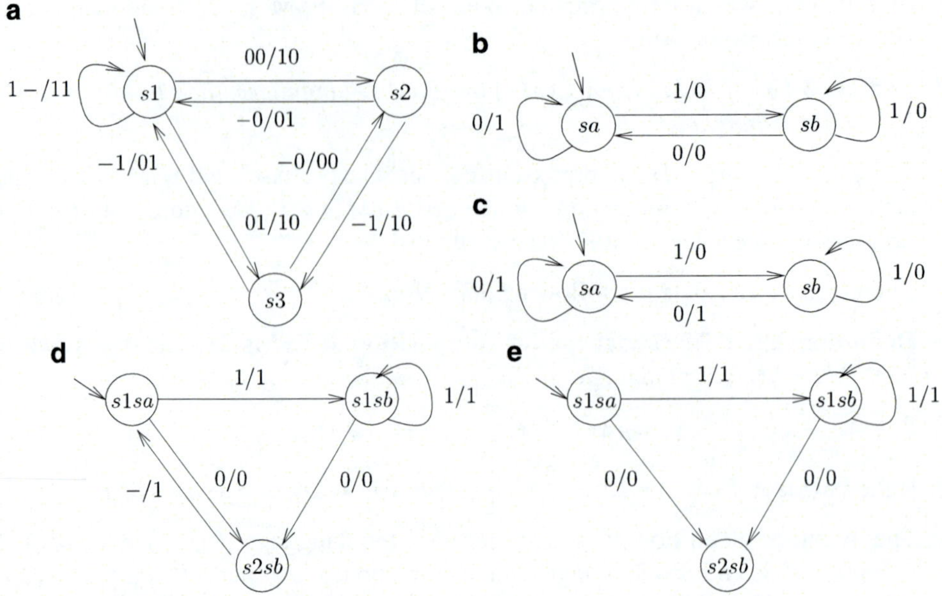

Fig. 3.1 Illustration of synchronous compositions $M_A \bullet M_B = M_{A\bullet B}$ and $M_A \bullet M'_B = M_{A\bullet B'}$ of Example 3.11-a) and 3.11-b). **(a)** FSM M_A; **(b)** FSM M_B; **(c)** FSM M'_B; **(d)** FSM $M_A \bullet M_B = M_{A\bullet B}$; **(e)** FSM $M_A \bullet M'_B = M_{A\bullet B'}$

$(0,\ (s1, sb),\ (s2, sb),\ 0)\}$ defines a partial FSM (no transition from state $(s2, sb)$).

Figure 3.1 shows FSMs M_A, M'_B and $M_A \bullet M'_B$.

Theorem 3.12. *Let M_A be a complete FSM over input alphabet $I_1 \times V$ and output alphabet $O_1 \times U$ and let M_B be a complete Moore FSM over input alphabet $I_2 \times U$ and output alphabet $O_2 \times V$. Then the composition $M_A \bullet M_B$ is a complete FSM.*

Proof. Consider a string $\alpha \in L(M_A)_{\uparrow I_2 \times O_2} \cap L(M_B)_{\uparrow I_1 \times O_1}$. Suppose that from the initial state, M_A reaches state s under the string $\alpha_{\downarrow I_1 \times U \times V \times O_1}$ and similarly that M_B reaches state t under the string $\alpha_{\downarrow I_2 \times U \times V \times O_2}$. Let the external input $(i_1, i_2) \in I_1 \times I_2$ be applied next. For any $u \in U$ there is a transition $i_2u, t, t', o_2v)$ in M_B, because M_B is a complete FSM; similarly, for any $v \in V$ there is a transition (i_1v, s, s', o_1u') in M_A, because M_A is a complete FSM. Moreover, given the input i_2u', there is a transition (i_2u', t, t'', o_2v) with the same output o_2v of transition (i_2u, t, t', o_2v), because M_B is a Moore FSM. Therefore u' and v are matching internal signals, i.e., the string α can be extended by $(i_1, i_2, u', v, o_1, o_2)$. □

3.2.2 *Parallel Composition of FSMs*

Consider the pair of FSMs[6,7]

1. FSM M_A has input alphabet $I_1 \cup V$, output alphabet $U \cup O_1$, and transition relation T_A.
2. FSM M_B has input alphabet $I_2 \cup U$, output alphabet $V \cup O_2$, and transition relation T_B.

We define a parallel composition operator $\diamond$ that associates a pair of FSMs M_A and M_B with another FSM $M_A \diamond M_B$ such that:

1. The alphabet of the external inputs is $I_1 \cup I_2 = I$.
2. The alphabet of the external outputs is $O_1 \cup O_2 = O$.

Recall that, by definition of parallel composition of languages, a sequence $\alpha \in ((I_1 \cup I_2)(O_1 \cup O_2))^\star$ is in the language of the parallel composition of $L(M_A)$ and $L(M_B)$ iff α is in the restriction onto $I_1 \cup I_2 \cup O_1 \cup O_2$ of the intersection of the expansion of $L(M_A)$ over $I_2 \cup O_2$ and of the expansion of $L(M_B)$ over $I_1 \cup O_1$:

$$\alpha \in L(M_A) \diamond L(M_B) \text{ iff}$$

$$\alpha \in [L(M_A)_{\Uparrow I_2 \cup O_2} L \cap L(M_B)_{\Uparrow I_1 \cup O_1}]_{\Downarrow I \cup O}.$$

Notice that the expansions $L(M_A)_{\Uparrow I_2 \cup O_2}$ and $L(M_B)_{\Uparrow I_1 \cup O_1}$ are needed to have the languages of M_A and M_B defined on the same alphabet; e.g., $L(M_B)$ is defined over $I_2 \cup U \cup V \cup O_2$, and the expansion $\Uparrow I_1 \times O_1$ defines it over $I_1 \cup I_2 \cup U \cup V \cup O_1 \cup O_2$. In this way we describe the behavior of each FSM as a component of the parallel composition.

Lemma 3.13. *If $L(M_A)$ and $L(M_B)$ are FSM $\cup$-languages, then $L(M_A) \diamond L(M_B) \cap (IO)^\star$ is an FSM $\cup$-language.*

Proof. $L(M_A) \diamond L(M_B) \cap (IO)^\star$ is IO-prefix-closed, because IO-prefix-closed $\cup$-languages are closed under $\diamond$ composition. Indeed, a state of the finite automaton corresponding to an FSM $\cup$-language is accepting iff it is the initial state or all its ingoing edges are labeled by symbols in O. The property is preserved by intersection and restriction over $I \cup O$. The intersection with $(IO)^\star$ makes sure that in the strings of the resulting FSM $\cup$-language an input is always followed by exactly one output, so that a corresponding FSM (with edges labeled by pairs (i/o)) can be reconstructed. Notice that $L(M_A) \diamond L(M_B) \cap (IO)^\star$ does not need to be IO-progressive, because partial FSMs are allowed. □

---

[6] Notice that more complex topologies can be handled in the same way, by defining the compositions with respect to the appropriate sets of variables, as remarked for synchronous composition.

[7] For simplicity the alphabets I_1, I_2, O_1, O_2, U, V are assumed to be pairwise disjoint.

Therefore we can state the following definition.

Definition 3.2.2. The **parallel composition of FSMs** M_A and M_B yields the FSM $M_A \diamond M_B$ with language

$$L(M_A \diamond M_B) = L(M_A) \diamond L(M_B) \cap (IO)^\star.$$

If the language $L(M_A) \diamond L(M_B) \cap (IO)^\star = \{\epsilon\}$, then $M_A \diamond M_B$ is a trivial FSM.

The previous definition is sound because the language $L(M_A) \diamond L(M_B) \cap (IO)^\star$ by Lemma 3.13 is an FSM language, which corresponds to a (partial) complete FSM if the language $L(M_A) \diamond L(M_B) \cap (IO)^\star$ is (not) IO-progressive. Then by subset construction and state minimization we produce a reduced observable FSM.

Example 3.14. (a) Parallel composition of two FSMs defining a complete FSM.

Given the topology shown in Fig. 1.1d, consider the FSMs $M_A = \langle S_A, I_1 \cup V, U \cup O_1, T_A, s1\rangle$ and $M_B = \langle S_B, U, V, T_B, sa\rangle$ with $S_A = \{sa, sb\}$, $T_A = \{(i_1,\ sa,\ sa,\ o_1),\ (v_1,\ sa,\ sa,\ o_1),\ (v_2,\ sa,\ sa,\ u_1),\ (i_2,\ sa,\ sb,\ u_2),\ (v_1,\ sb,\ sb,\ u_1),\ (v_2,\ sb,\ sb,\ o_2),\ (i_1,\ sb,\ sa,\ o_1),\ (i_2,\ sb,\ sa,\ o_1)\}$, $S_B = \{s1, s2\}$, $T_B = \{(u_2,\ s1,\ s2,\ v_2),\ (u_1\ s1,\ s2,\ v_1),\ (u_1,\ s2,\ s2,\ v_2),\ (u_2,\ s2,\ s1,\ v_2)\}$.

Then the composition $M_A \diamond M_B = M_{A\diamond B} = \langle S_{A\diamond B}, I_1, O_1, T_{A\diamond B}, (sa, s1)\rangle$ with $S_{A\diamond B} = \{(sa, s1), (sb, s2), (sa, s2), (sb, s1)\}$ and $T_{A\diamond B} = \{(i_1,\ (sa, s1),\ (sa, s1),\ o_1),\ (i_2,\ (sa, s1),\ (sb, s2),\ o_2),\ (i_1,\ (sb, s2),\ (sa, s2),\ o_1),\ (i_2,\ (sb, s2),\ (sa, s2),\ o_1),\ (i_1,\ (sa, s2),\ (sa, s2),\ o_1),\ (i_2,\ (sa, s2),\ (sb, s1),\ o_2),\ (i_1,\ (sb, s1),\ (sa, s1),\ o_1),\ (i_2,\ (sb, s1),\ (sa, s1),\ o_1)\}$ is a complete FSM.

Figure 3.2 shows some steps and the result of the computation of $M_A \diamond M_B$.

(b) Parallel composition of two FSMs defining a partial FSM.

Modify the transition relation of M_B as follows: $T_B = \{(u_2,\ s1,\ s2,\ v_2),\ (u_1\ s1,\ s2,\ v_1),\ (u_1,\ s2,\ s2,\ v_1),\ (u_2,\ s2,\ s1,\ v_1)\}$. Then $T_{A\diamond B} = \{(i_1,\ (sa, s1),\ (sa, s1),\ o_1),\ (i_2,\ (sa, s1),\ (sb, s2),\ o_2),\ (i_1,\ (sb, s2),\ (sa, s2),\ o_1),\ (i_2,\ (sb, s2),\ (sa, s2),\ o_1),\ (i_1,\ (sa, s2),\ (sa, s2),\ o_1)\}$ defines a partial FSM, because there is no transition from state $(sa, s2)$ under input i_2 (from $(sa, s2)$ under i_2/u_2 we reach $(sb, s2)$, then under u_2/v_1 we reach $(sb, s1)$, then under v_1/u_1 we stay in $(sb, s1)$, then under u_1/v_1 we reach $(sb, s2)$, where there is an endless cycle – livelock – under v_1/u_1 to $(sb, s2)$ and then under u_1/v_1 back again to $(sb, s2)$, without any production of an output in O_1).

3.2.3 *Equations Over FSMs*

Consider the network of FSMs shown in Fig. 3.3, where FSM M_A has input signals I_1 and V and output signals U and O_1, and FSM M_X has input signals I_2 and U and

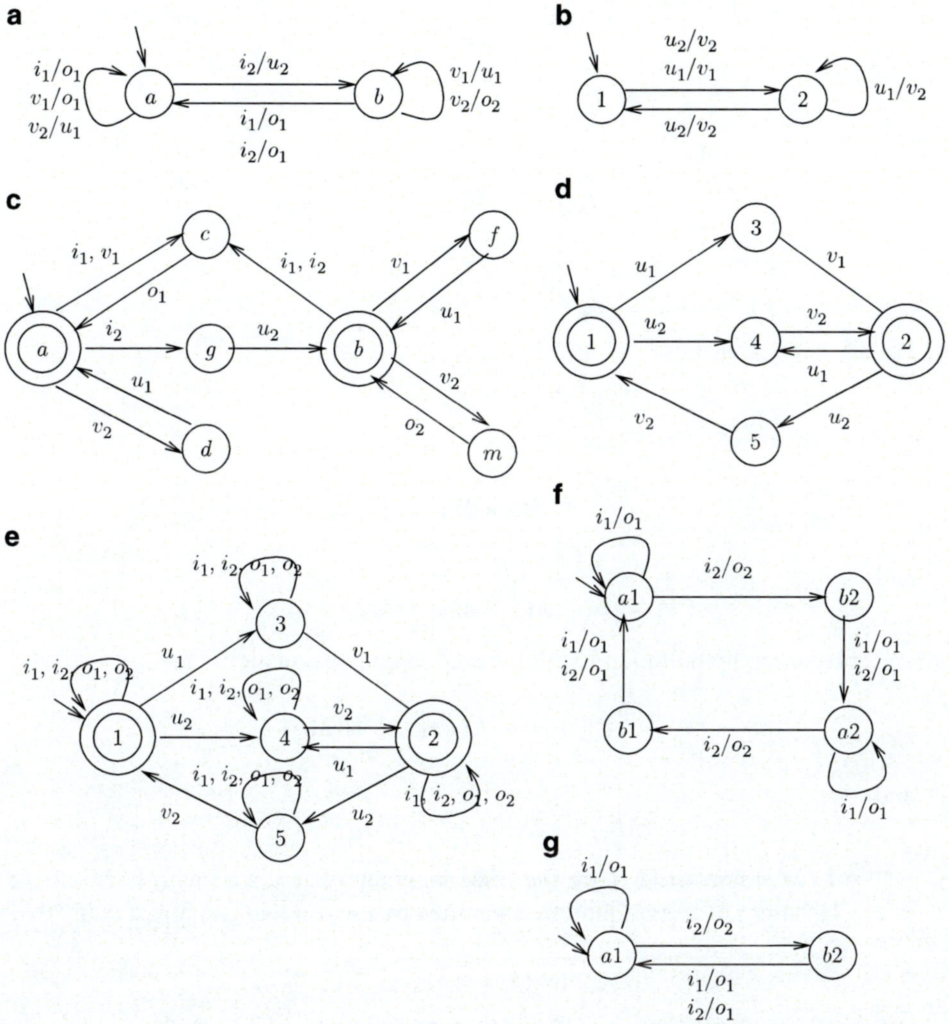

Fig. 3.2 Illustration of parallel composition $M_A \diamond M_B = M_{A\diamond B}$ of Example 3.14-a. **(a)** FSM M_A; **(b)** FSM M_B; **(c)** Automaton of A ($\cup$-language of M_A); **(d)** Automaton of B ($\cup$-language of M_B); **(e)** Automaton of $B_{\Uparrow I_1 \cup O_1}$; **(f)** FSM $M_A \diamond M_B = M_{A\diamond B}$; **(g)** Reduced FSM $M_A \diamond M_B = M_{A\diamond B}$. *Double-circled* nodes represent accepting states in (c), (d), (e). In pictures of FSMs, all nodes are accepting

output signals V and O_2. The network implements a specification M_C with input signals I_1, I_2 and output signals O_1, O_2. Supposing that M_A and M_C are known and M_X is unknown, we want to define an equation of the type $M_A \odot M_X \approx M_C$, to capture the FSMs M_B that in place of M_X let the network of M_A and M_B match the specification M_C. Through Definitions 3.1.7 and 3.1.8 we have seen two different ways to associate an FSM language with a given FSM, and related composition operators $\bullet$ and $\diamond$ have been introduced in Sect. 3.2; therefore we introduce two

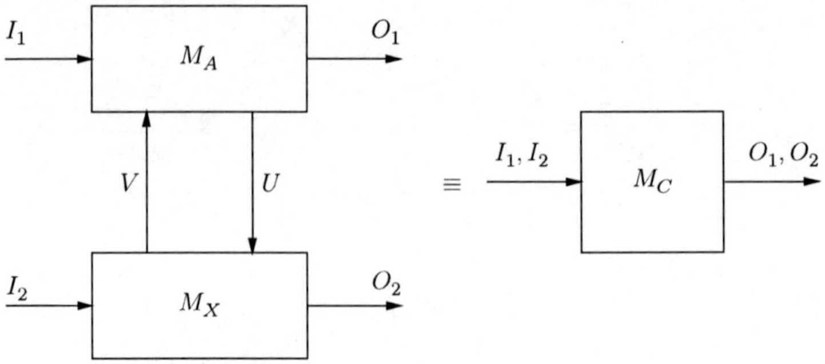

Fig. 3.3 General topology

types of equations over FSMs:

$$M_A \bullet M_X \preceq M_C$$

and

$$M_A \diamond M_X \preceq M_C,$$

and solve them by building first the related language equations

$$L(M_A) \bullet L(M_X) \subseteq L(M_C)$$

and

$$L(M_A) \diamond L(M_X) \subseteq L(M_C) \cup \overline{(IO)^\star},$$

where $L(M_A)$ and $L(M_C)$ are the FSM languages associated with FSMs M_A and M_C. The latter language equation is justified by the following chain of equivalences

$$\begin{aligned} M_A \diamond M_X \preceq M_C &\Leftrightarrow \\ L(M_A \diamond M_X) \subseteq L(M_C) &\Leftrightarrow \textit{by Def. 3.2.2} \\ L(M_A) \diamond L(M_X) \cap (IO)^\star \subseteq L(M_C) &\Leftrightarrow \\ L(M_A) \diamond L(M_X) \subseteq L(M_C) \cup \overline{(IO)^\star}. \end{aligned}$$

The last equivalence uses the set-theoretic equality $A \cap B \subseteq C \Leftrightarrow A \subseteq C + \overline{B}$.[8]

When there is no ambiguity we will denote by $A \bullet X \subseteq C$ and $A \diamond X \subseteq C \cup \overline{(IO)^\star}$ the language equations $L(M_A) \bullet L(M_X) \subseteq L(M_C)$ and $L(M_A) \diamond L(M_X) \subseteq L(M_C) \cup \overline{(IO)^\star}$, where $L(M_A)$, $L(M_X)$ and $L(M_C)$ are, respectively, the $\times$-languages and $\cup$-languages associated with the FSMs M_A, M_X and M_C.

---

[8]In one direction, $A \cap B \subseteq C \Rightarrow A \cap B + A \cap \overline{B} \subseteq C + A \cap \overline{B} \Rightarrow A \subseteq C + A \cap \overline{B} \Rightarrow A \subseteq C + \overline{B}$. In the other direction, $A \subseteq C + \overline{B} \Rightarrow A \cap B + A \cap \overline{B} \subseteq C + \overline{B} \Rightarrow A \cap B \subseteq C$, because $A \cap B \not\subseteq \overline{B}$ if $A \cap B \neq \emptyset$ (if $A \cap B = \emptyset$ then $A \cap B \subseteq C$).

3.3 Solution of FSM Equations Under Synchronous Composition

3.3.1 Largest FSM Solution

Given alphabets I_1, I_2, U, V, O_1, O_2, an FSM M_A over inputs $I_1 \times V$ and outputs $U \times O_1$, and an FSM M_C over inputs $I_1 \times I_2$ and outputs $O_1 \times O_2$, consider the FSM equations

$$M_A \bullet M_X \preceq M_C, \tag{3.1}$$

or,

$$M_A \bullet M_X \cong M_C, \tag{3.2}$$

whose unknown is an FSM M_X over inputs $I_2 \times U$ and outputs $V \times O_2$. Sometimes the shortened notation $I = I_1 \times I_2$ and $O = O_1 \times O_2$ will be used.[9] A variant of equations is

Definition 3.3.1. FSM M_B is a **solution** of the equation $M_A \bullet M_X \preceq M_C$, where M_A and M_C are FSMs, iff $M_A \bullet M_B \preceq M_C$.

FSM M_B is a **solution** of the equation $M_A \bullet M_X \cong M_C$, where M_A and M_C are FSMs, iff $M_A \bullet M_B \cong M_C$.

Converting to the related FSM languages, we construct the associated language equation

$$L(M_A) \bullet L(M_X) \subseteq L(M_C), \tag{3.3}$$

where $L(M_A)$ is an FSM language over alphabet $I_1 \times U \times V \times O_1$, $L(M_C)$ is an FSM language over alphabet $I_1 \times I_2 \times O_1 \times O_2$ and the unknown FSM language is over alphabet $I_2 \times U \times V \times O_2$. The previous equation can be rewritten for simplicity as

$$A \bullet X \subseteq C. \tag{3.4}$$

We want to characterize the solutions of $A \bullet X \subseteq C$ as FSM languages. We know from Theorem 2.12 that the largest solution of the equation $A \bullet X \subseteq C$ is the language $S = \overline{A \bullet \overline{C}}$.

When A and C are FSM languages, the following property holds.

Theorem 3.15. *Given the equation $A \bullet X \subseteq C$ and its largest solution $S = \overline{A \bullet \overline{C}}$, then $S \neq \emptyset$.*

Proof. FSM languages are not closed under complementation, because the complement of an FSM language does not include the empty string ϵ, so it cannot be an FSM language. So $\overline{C}$ does not include ϵ and neither does $\overline{C}_{\uparrow U \times V}$. Then the

[9] Equations over more complex topologies and with respect to other input and output alphabets can be solved similarly, by setting up the appropriate compositions, as pointed out already when introducing composition of automata and FSMs.

intersection $A_{\uparrow I_2 \times O_2} \cap \overline{C}_{\uparrow U \times V}$ does not include the empty string, neither does its projection $(A_{\uparrow I_2 \times O_2} \cap \overline{C}_{\uparrow U \times V})_{\downarrow I_2 \times U \times V \times O_2}$, that is $A \bullet \overline{C}$. Therefore $\overline{A \bullet \overline{C}}$ includes the empty string, i.e., $\epsilon \in S \neq \emptyset$. □

Example 3.16. Consider the FSMs $M_A = \langle S_A, I_1 \times V, U \times O_1, T_A, sa \rangle$ and $M_C = \langle S_C, U, V, T_C, s1 \rangle$ with $S_A = \{sa\}$, $T_A = \{(01, sa, sa, 01), (00, sa, sa, 01), (11, sa, sa, 10), (10, sa, sa, 10)\}$, $S_C = \{s1\}$, $T_C = \{(1, s1, s1, 1), (0, s1, s1, 0)\}$. The equation $M_A \bullet M_X \preceq M_C$ yields the language equation $A \bullet X \subseteq C$ with solution $S = \{\epsilon\}$, i.e., the corresponding FSM solution M_X produces only the empty word. The reason is that the i/o combinations of M_C, i.e., $1/1, 0/0$ do not match any i/o combination of M_A, i.e., $0/1, 1/0$.

In general S is not an FSM language. To compute the largest FSM language contained in S, that is S^{FSM}, we must compute the largest prefix-closed language contained in S.

Theorem 3.17. *Let A and C be FSM languages. The largest FSM language that is a solution of the equation $A \bullet X \subseteq C$ is given by S^{FSM}, where $S = \overline{A \bullet \overline{C}}$. S^{FSM} is obtained by applying Procedure 3.1.1 to S. S^{FSM} contains at least the string ϵ.*

If $A \bullet S^{FSM} = C$ then S^{FSM} is the largest solution of the FSM language equation $A \bullet X = C$.

Thus, the synchronous FSM language equation $A \bullet X \subseteq C$ is always solvable, since the solution includes at least the empty string and so its prefix-closure does too. In other words, the trivial FSM is always a solution of the synchronous FSM equation $A \bullet X \subseteq C$.

Example 3.18. Consider the equation $M_A \bullet M_X \preceq M_C$, with M_A and M_C shown, respectively, in Fig. 3.4a,c. The FSMs are composed with respect to the rectification topology and their alphabets are $I_1 = \{i_1, i_2\}$, $O_1 = \{o_1, o_2\}$, $U = \{u_1, u_2\}$ and $V = \{v_1, v_2\}$. M_A is a partial FSM, e.g., the transitions from state a under inputs $i_1 v_2$ and $i_2 v_2$ are missing. The automata of the related languages are shown, respectively, in Fig. 3.4b,d. The intermediate steps to compute the solution are demonstrated in Fig. 3.4e–g. The automaton generating the largest language solution, $S = \overline{(A \cap (\overline{C}_{\uparrow U \cup V})_{\downarrow U \cup V}}$, is portrayed in Fig. 3.4h. The state labeled dc accepts strings that are not in the language of the context. Notice that S is not prefix-closed, since string $u_1 v_1 u_2 v_1 \in S$, but $u_1 v_1 \notin S$; its largest prefix-closed sublanguage yields the largest FSM solution M_X shown in Fig. 3.4i. M_X is a PNDFSM and the composition of any DFSM reduction of M_X with M_A produces the trivial machine.

Example 3.19. Consider the following variants of the context FSM M_A in the equation $M_A \bullet M_X \preceq M_C$, studied in Example 3.18. The specification FSM M_C is the same. The variants of M_A are M_A^1, M_A^2 and M_A^3, shown respectively in Fig. 3.5a,d,f. They are all complete FSMs, whereas M_A of Example 3.18 is a partial FSM. The solution of $M_A^1 \bullet M_X \preceq M_C$ is M_X^1 in Fig. 3.5c. M_X^1 is a partial FSM and $M_A^1 \bullet M_X^1 = M_\epsilon$, i.e., the composition yields the trivial FSM that is the "ultimate" partial machine, because each external input sequence causes a violation

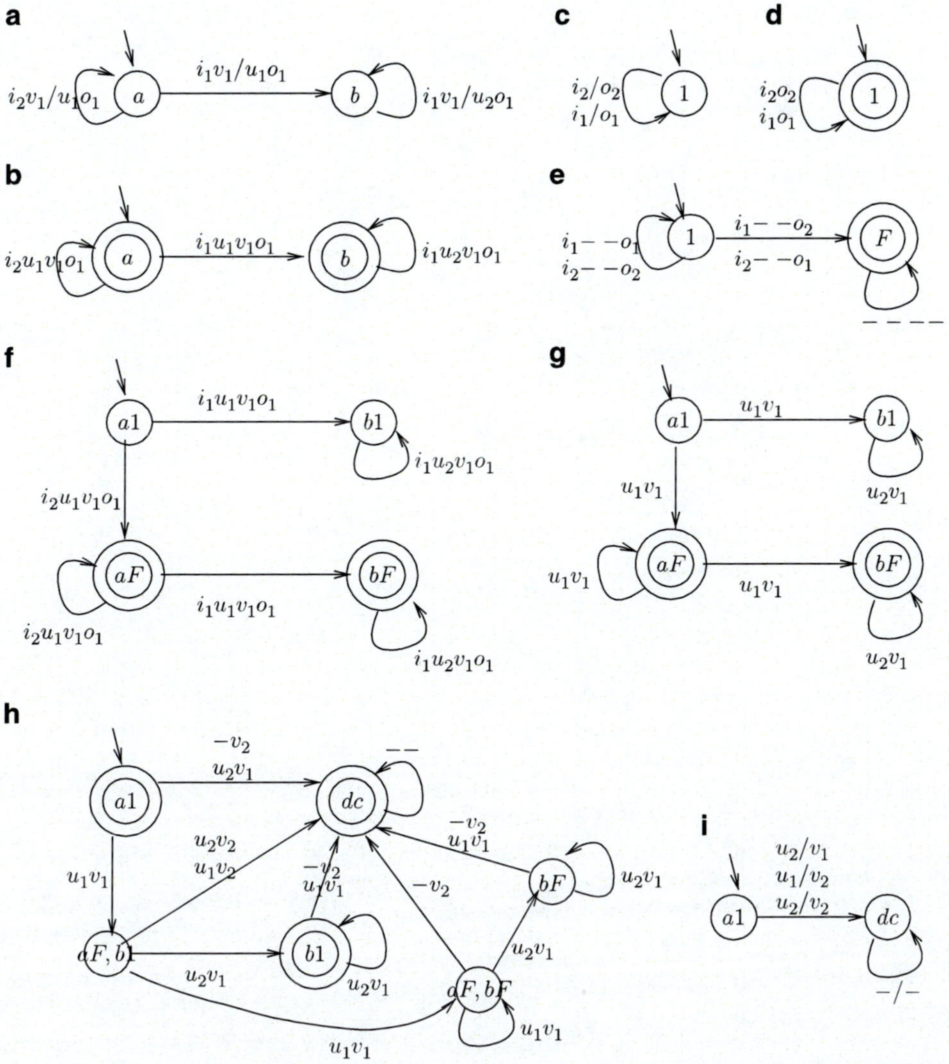

Fig. 3.4 Illustration of Example 3.18. (**a**) FSM M_A; (**b**) FA (finite automaton) of $A = L_r^{\times}(M_A)$; (**c**) FSM M_C; (**d**) FA of $C = L_r^{\times}(M_C)$; (**e**) FA of $\overline{C}_{\uparrow U \times V}$; (**f**) FA of $A \cap \overline{C}_{\uparrow U \times V}$; (**g**) FA of $(A \cap \overline{C}_{\uparrow U \times V})_{\downarrow U \times V}$; (**h**) FA of largest solution $S = \overline{(A \cap \overline{C}_{\uparrow U \times V})_{\downarrow U \times V}}$; (**i**) Largest FSM solution M_X. M_X is a PNDFSM such that $M_A \bullet M_X = M_\epsilon \preceq M_C$

of the specification. Adding to state a of M_A^1 the transition to itself i_1v_1/u_2o_2 would produce an FSM $M_A^{1'}$ yielding the solution $M_X^{1'} = M_\epsilon$. The solution of $M_A^2 \bullet M_X \preceq M_C$ is M_X^2 in Fig. 3.5e. M_X^2 is a complete FSM and $M_A^2 \bullet M_X^2 \preceq M_C$, where $M_A^2 \bullet M_X^2$ is an incomplete FSM that is a reduction of M_C. The solution of $M_A^3 \bullet M_X \preceq M_C$ is M_X^3 in Fig. 3.5g. M_X^3 is a complete FSM and $M_A^3 \bullet M_X^3 \cong M_C$ is a complete FSM.

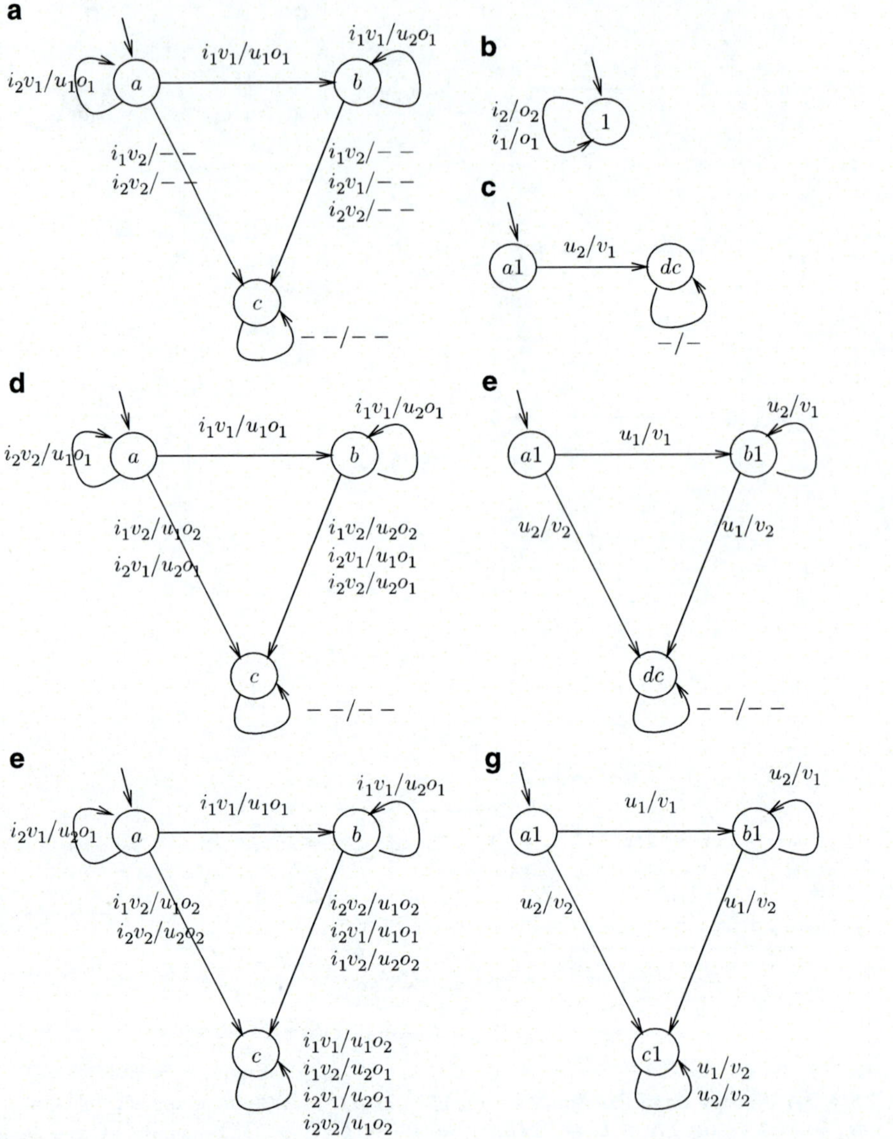

Fig. 3.5 Illustration of Example 3.19. **(a)** FSM M_A^1; **(b)** FSM M_C; **(c)** FSM M_X^1; **(d)** FSM M_A^2; **(e)** FSM M_X^2; **(f)** FSM M_A^3; **(g)** FSM M_X^3

For logic synthesis applications, we assume that M_A and M_C are complete FSMs and we require that the solution is a complete FSM too. This is obtained by applying Procedure 3.1.2 to S^{FSM}, yielding $Prog(S^{FSM})$, the largest $(I_2 \times U)$-progressive FSM language $\subseteq (I_2 \times U \times V \times O)^\star$. Notice that an $(I_2 \times U)$-progressive solution might not exist, and in that case Procedure 3.1.2 returns an empty language.

Proposition 3.20. *FSM M_B is a solution of the equation $M_A \bullet M_X \preceq M_C$, where M_A and M_C are FSMs, iff M_B is a reduction of the FSM M_S associated to S^{FSM}, where S^{FSM} is obtained by applying Procedure 3.1.1 to S, where $S = \overline{A \bullet \overline{C}}$. If $S^{FSM} = \{\epsilon\}$ then the trivial FSM is the only solution.*

The largest complete FSM solution $M_{Prog(S^{FSM})}$ is found, if it exists, by applying Procedure 3.1.2.

A complete FSM is a solution of $M_A \bullet M_X \preceq M_C$ iff it is a reduction of the largest complete solution $M_{Prog(S^{FSM})}$.

Given the largest (complete) solution $M_{Prog(S^{FSM})}$ of the equation $M_A \bullet M_X \preceq M_C$, let $M_A \bullet M_X \cong M_C$. Then $M_{Prog(S^{FSM})}$ is the largest (complete) solution of the equation $M_A \bullet M_X \cong M_C$.

Example 3.21. (Variant of Example 3.16) Consider the FSMs $M_A = \langle S_A, I_1 \times V, U \times O_1, T_A, sa\rangle$ and $M_C = \langle S_C, U, V, T_C, s1\rangle$ with $S_A = \{sa\}$, $T_A = \{(01,\ sa,\ sa,\ s00),\ (00,\ sa,\ sa,\ 01),\ (11,\ sa,\ sa,\ 10),\ (10,\ sa,\ sa,\ 10)\}$, $S_C = \{s1\}$, $T_C = \{(1,\ s1,\ s1,\ 1),\ (0,\ s1,\ s1,\ 0)\}$. The equation $M_A \bullet M_X \preceq M_C$ yields the language equation $A \bullet X \subseteq C$ with solution $S = \{(01)^\star\}$, i.e., the corresponding FSM solution M_X produces the set of strings of input/output pairs $\{(0/1)^\star\}$, and so the equation has no complete FSM solution.

Not every deterministic complete FSM M_{SR} that is a reduction of the largest solution M_S is a solution of the equation $M_A \bullet M_X \cong M_C$, as shown by the examples in Example 3.22.

Example 3.22. Given the context FSM M_A and the specification FSM M_C shown respectively in Fig. 3.6a,b, consider the equation $M_A \bullet M_X \cong M_C$, whose largest solution M_S is in Fig. 3.6c (and obviously satisfies $M_A \bullet M_X \cong M_C$, as shown in Fig. 3.6d). Then the deterministic complete FSM M_{S1} in Fig. 3.6e is not a solution of $M_A \bullet M_X \cong M_C$, whereas M_{S2} in Fig. 3.6g is a solution of $M_A \bullet M_X \cong M_C$, as shown respectively in Fig. 3.6f and Fig. 3.6h. Notice that both M_{S1} and M_{S2} are reductions of the largest solution M_S.

Given the context FSM M_A and the specification FSM M_C shown respectively in Fig. 3.7a,b, consider the equation $M_A \bullet M_X \cong M_C$, whose largest solution M_S is in Fig. 3.7c (and obviously satisfies $M_A \bullet M_X \cong M_C$, as shown in Fig. 3.7d). Then a deterministic complete FSM that is a reduction of M_S should choose at state s either the transition u_3/v_2 or u_3/v_1. Suppose to define the reduction M_{S1} where at state s we choose u_3/v_2 instead of u_3/v_1, then it is $M_A \bullet M_{S1} \not\cong M_C$, because in the product at state 1 we cannot get the transition i_1/o_1 that requires the selection u_3/v_1 in M_{S1}; similarly, if we define the reduction M_{S2} where at state s we choose u_3/v_1 instead of u_3/v_2. then it is $M_A \bullet M_{S1} \not\cong M_C$, because in the product at state 1 we cannot get the transition i_3/o_2 that requires the selection u_3/v_2 in M_{S2}. So there is no deterministic complete FSM that is a reduction of the largest solution M_S and also solves the FSM equation $M_A \bullet M_X \cong M_C$.

Another interesting example and related discussion can be found in Example 3.1 and Fig. 2 of [149], where by removing any transition from the largest solution M_S we would miss some complete deterministic FSM reduction that is a solution.

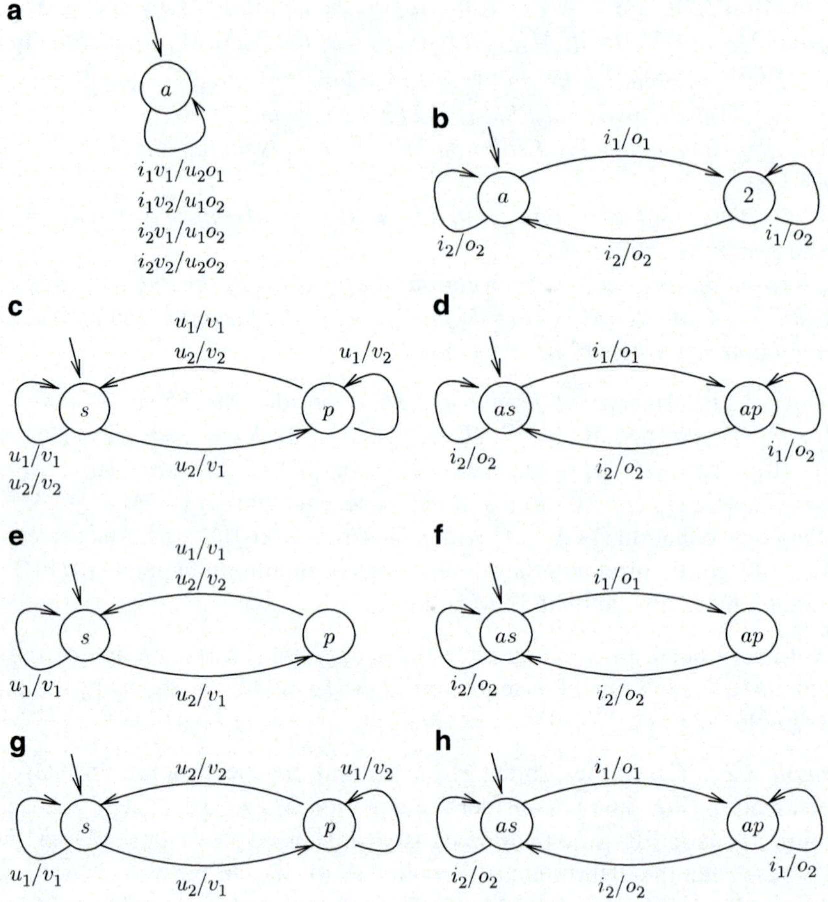

Fig. 3.6 Illustration of Example 3.18. (**a**) Context FSM M_A; (**b**) Specification FSM M_C; (**c**) Largest Solution FSM M_S; (**d**) $M_A \bullet M_S \cong M_C$; (**e**) FSM M_{S1}, reduction of M_S; (**f**) $M_A \bullet M_{S1} \ncong M_C$; (**g**) FSM M_{S2}, reduction of M_S; (**h**) $M_A \bullet M_{S2} \cong M_C$

3.3.1.1 Complexity of the Algorithm

Consider the equation $A \bullet X \subseteq C$, where A and C are FSM languages. We know that the largest unconstrained solution is given by $S = \overline{A \bullet \overline{C}}$. Given the rectification topology in Fig. 1.1d with M_B as the unknown M_X, the solution has in the worst-case $2^{|S_A|.2^{|S_C|}}$ states, where S_A are the states of FA A and S_C are the states of FA C. The exponent $2^{|S_C|}$ appears when C is non-deterministic, to account for the determinization needed to compute $\overline{C}$; otherwise if C is deterministic, in place of $2^{|S_C|}$ we have $|S_C|$, because complementation becomes a linear operation. The product $|S_A|.2^{|S_C|}$ is due to the product of automata A and C. Then to complete the computation of the $\bullet$ operator we must project the product on the internal signals U and V; as a result we may get again a non-deterministic automaton, and therefore

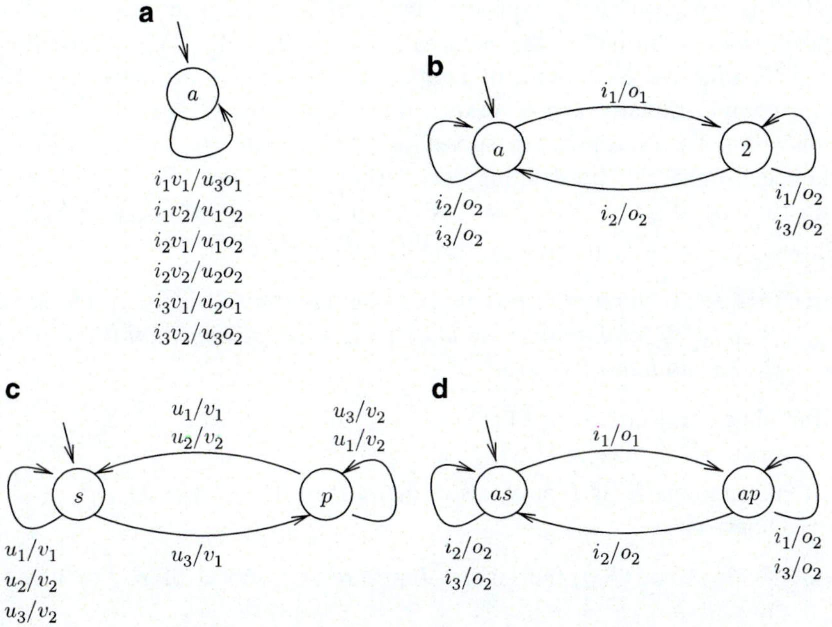

Fig. 3.7 Illustration of Example 3.18. **(a)** Context FSM M_A; **(b)** Specification FSM M_C; **(c)** Largest Solution FSM M_S; **(d)** $M_A \bullet M_S \cong M_C$

a new determinization is needed before performing the final complementation: this explains the outer exponential $2^{|S_A|.2^{|S_C|}}$.

There are "easier" topologies, like supervisory control, where there is no projection onto a subset of signals of the product automaton; therefore non-determinism is not introduced and so the final complementation is linear, resulting in the complexity of $|S_A|.2^{|S_C|}$. Moreover, if S_C is deterministic, the exponential $2^{|S_C|}$ is replaced by $|S_C|$, and so the final complexity of supervisory control is bounded by only $|S_A|.|S_C|$.

The operations on the language S to extract from it an FSM language, a complete FSM language or a Moore solution (see below) are linear in the number of edges of the automaton representing S.

3.3.2 Restricted FSM Compositional Solutions

It is interesting to compute the subset of compositionally I-progressive solutions B, i.e., such that $A_{\uparrow I_2 \times O_2} \cap B_{\uparrow I_1 \times O_1}$ is an I-progressive FSM language $\subseteq (I \times U \times V \times O)^{\star}$. Thus the composition (after projection on the external signals) is the language of a complete FSM over inputs $I_1 \times I_2$ and outputs $O_1 \times O_2$. Since $A_{\uparrow I_2 \times O_2} \cap S^{FSM}_{\uparrow I_1 \times O_1}$ is prefix-closed and hence corresponds to a partial FSM, we have to restrict it so that it is also I-progressive, which corresponds to a complete FSM.

If S^{FSM} is compositionally I-progressive, then S^{FSM} is the largest compositionally I-progressive solution of the equation. However, not every non-empty subset of S^{FSM} inherits the feature of being compositionally I-progressive. If S^{FSM} is not compositionally I-progressive, then denote by $cProg(S^{FSM})$ the largest compositionally I-progressive subset of S^{FSM}. Conceptually $cProg(S^{FSM})$ is obtained from S^{FSM} by deleting each string α such that, for some $i \in I$, there is no $(u, v, o) \in U \times V \times O$ such that $\alpha\ (i, u, v, o) \in A_{\uparrow I_2 \times O_2} \cap S^{FSM}_{\uparrow I_1 \times O_1}$. The following procedure tells how to compute $cProg(S^{FSM})$.

Procedure 3.3.1. *Input: Largest prefix-closed solution S^{FSM} of synchronous equation $A \bullet X \subseteq C$ and context A; Output: Largest compositionally I-progressive prefix-closed solution $cProg(S^{FSM})$.*

1. Initialize i to 1 and S^i to S^{FSM}.
2. Compute $R^i = A_{\uparrow I_2 \times O_2} \cap S^i_{\uparrow I_1 \times O_1}$.
   If the language R^i is I-progressive then $cProg(S^{FSM}) = S^i$.
   Otherwise
   (a) Obtain $Prog(R^i)$, the largest I-progressive subset of R^i, by using Procedure 3.1.2.
   (b) Compute $T^i = S^i \setminus (R^i \setminus Prog(R^i))_{\downarrow I_2 \times U \times V \times O_2}$.
3. If $T^{i\ FSM} = \emptyset$ then $cProg(S^{FSM}) = \emptyset$.
   Otherwise
   (a) Assign the language $T^{i\ FSM}$ to S^{i+1}.
   (b) Increment i by 1 and go to 2.

Theorem 3.23. *Procedure 3.3.1 returns the largest compositionally I-progressive (prefix-closed) solution, if it terminates.*

Theorem 3.24. *Procedure 3.3.1 terminates.*

The proofs can be found in [148].

A sufficient condition to insure that $A_{\uparrow I_2 \times O_2} \cap S^{FSM}_{\uparrow I_1 \times O_1}$ is an I-progressive FSM language is that $S^{FSM}_{\uparrow I_1 \times O_1}$ or $A_{\uparrow I_2 \times O_2}$ satisfy the Moore property (see Theorem 3.12 for a related statement proved for complete FSMs). If S^{FSM} is Moore then it is the largest Moore solution of the equation. However, not every non-empty subset of S^{FSM} inherits the feature of being Moore. If S^{FSM} is not Moore, then denote by $Moore(S^{FSM})$ the largest Moore subset of S^{FSM}. The set $Moore(S^{FSM})$ is obtained by deleting from S^{FSM} each string which causes S^{FSM} to fail the Moore property.

Proposition 3.25. *If $Moore(S^{FSM}) \neq \emptyset$, then it is the largest Moore solution of the equation $A \bullet X \subseteq C$. Otherwise the equation $A \bullet X \subseteq C$ has no Moore solution.*

To compute the largest Moore FSM that is a solution, it is sufficient to apply Procedure 3.1.5 to the FSM $M_{S^{FSM}}$ associated with S^{FSM}, as justified by Theorem 3.9. The result is the largest Moore FSM solution, of which every deterministic Moore solution is a reduction.

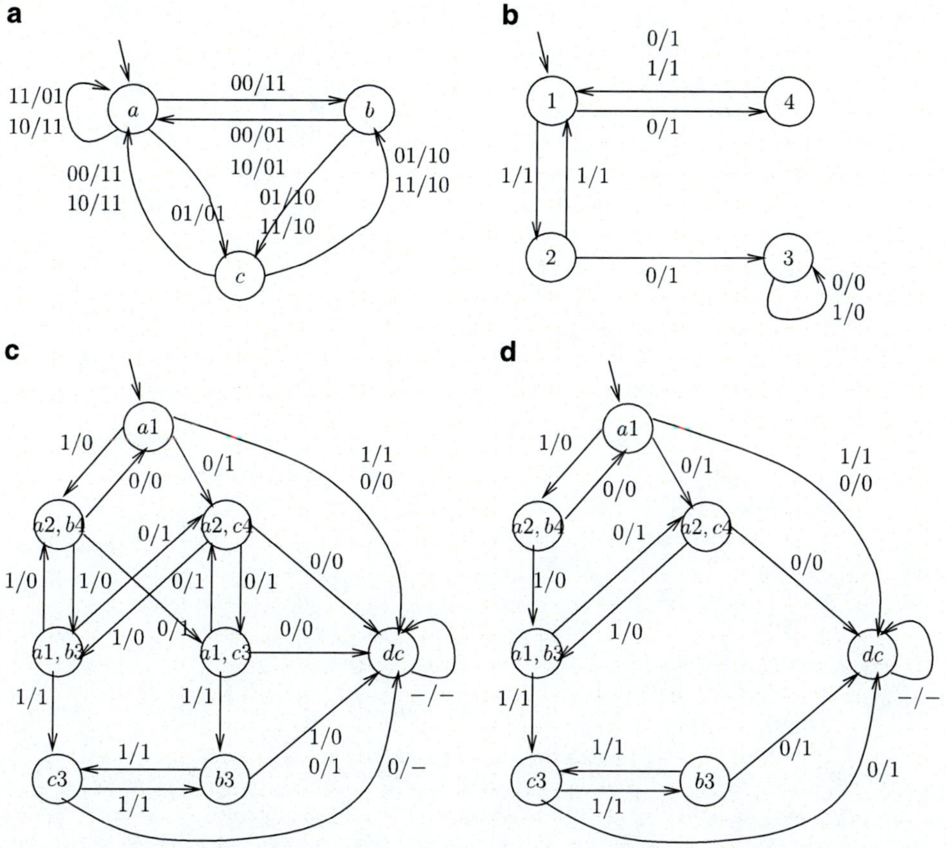

Fig. 3.8 Illustration of Example 3.26. (**a**) FSM M_A; (**b**) FSM M_C; (**c**) Largest FSM solution M_X; (**d**) Largest Moore FSM solution $Moore(M_X)$

Example 3.26. Consider the equation $M_A \bullet M_X \preceq M_C$, with M_A and M_C shown, respectively, in Fig. 3.8a,b. The largest FSM solution M_X is shown in Fig. 3.8c, whereas Fig. 3.8d shows the largest Moore FSM solution $Moore(M_X)$. Moore FSM solutions are portrayed in Fig. 3.9a,b, whereas nonMoore FSM solutions are pictured in Fig. 3.9c,d.

3.4 Solution of FSM Equations Under Parallel Composition

3.4.1 *Largest FSM Solution*

Given alphabets I_1, I_2, U, V, O_1, O_2, an FSM M_A over inputs $I_1 \cup V$ and outputs $U \cup O_1$, and an FSM M_C over inputs $I_1 \cup I_2$ and outputs $O_1 \cup O_2$, consider the FSM equations

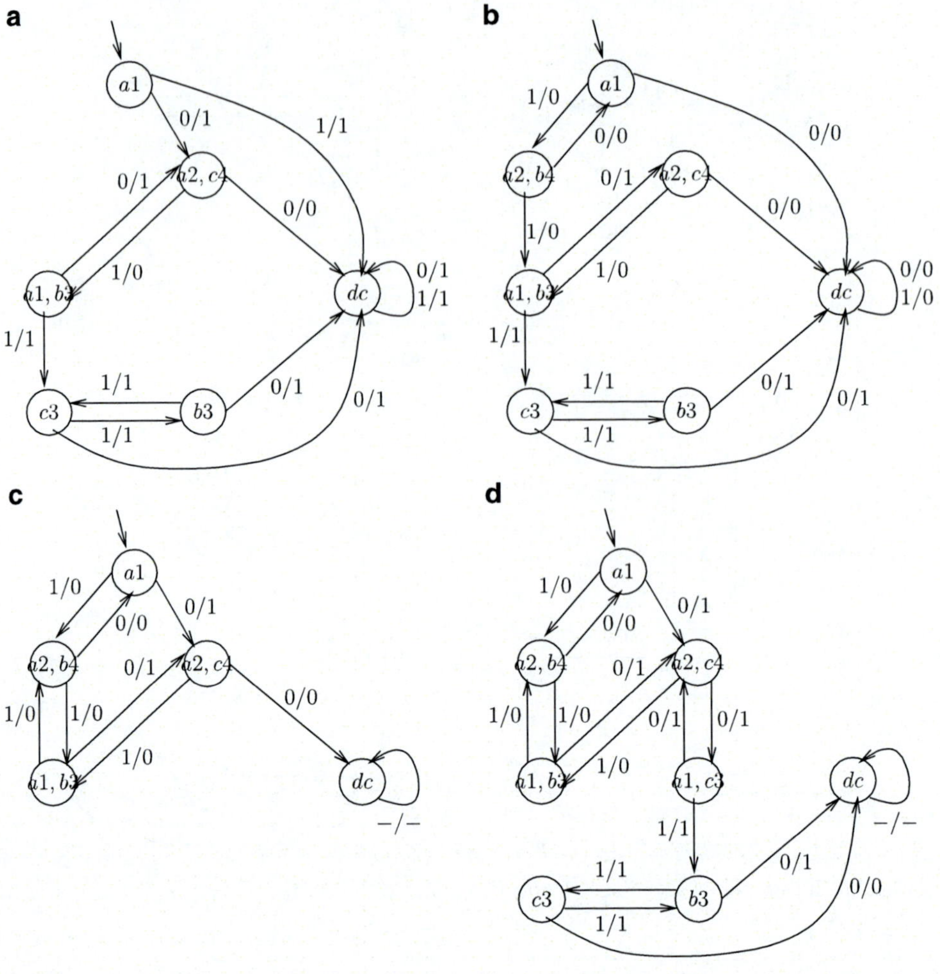

Fig. 3.9 Illustration of Example 3.26. **(a)–(b)** Moore FSM solutions; **(c)–(d)** Non-Moore FSM solutions

$$M_A \diamond M_X \preceq M_C, \tag{3.5}$$

or,

$$M_A \diamond M_X \cong M_C, \tag{3.6}$$

whose unknown is an FSM M_X over inputs $I_2 \cup U$ and outputs $V \cup O_2$. Sometimes the shortened notation $I = I_1 \cup I_2$ and $O = O_1 \cup O_2$ will be used.[10]

---

[10]Equations over more complex topologies and with respect to other input and output alphabets can be solved similarly, by setting up the appropriate compositions, as remarked already when discussing synchronous equations.

Definition 3.4.1. FSM M_B is a **solution** of the equation $M_A \diamond M_X \preceq M_C$, where M_A and M_C are FSMs, iff $M_A \diamond M_B \preceq M_C$.

FSM M_B is a **solution** of the equation $M_A \diamond M_X \cong M_C$, where M_A and M_C are FSMs, iff $M_A \diamond M_B \cong M_C$.

Converting to the related FSM languages, we construct the associated language equation (see Sect. 3.2.3)

$$L(M_A) \diamond L(M_X) \subseteq L(M_C) \cup \overline{(IO)^\star}, \tag{3.7}$$

where $L(M_A)$ is an FSM language over alphabet $I_1 \cup U \cup V \cup O_1$, $L(M_C)$ is an FSM language over alphabet $I_1 \cup I_2 \cup O_1 \cup O_2$ and the unknown FSM language is over alphabet $I_2 \cup U \cup V \cup O_2$. The previous equation can be rewritten for simplicity as

$$A \diamond X \subseteq C \cup \overline{(IO)^\star}. \tag{3.8}$$

We want to characterize the solutions of $A \diamond X \subseteq C \cup \overline{(IO)^\star}$ that are FSM languages. We know from Theorem 2.16 that the largest solution of the equation $A \diamond X \subseteq C \cup \overline{(IO)^\star}$ is the language $S = \overline{A \diamond (\overline{C} \cap (IO)^\star)}$.

In general S is not an FSM language. To compute the largest FSM language contained in S, that is S^{FSM}, we must compute the largest prefix-closed language contained in $S \cap ((I_2 \cup U)(V \cup O_2))^\star$.

Theorem 3.27. *Let A and C be FSM languages. The largest FSM language that is a solution of the equation $A \diamond X \subseteq C \cup \overline{(IO)^\star}$ is given by S^{FSM}, where $S = \overline{A \diamond (\overline{C} \cap (IO)^\star)}$. If $S = \emptyset$ then $S^{FSM} = \emptyset$; if $S \neq \emptyset$, S^{FSM} is obtained by applying Procedure 3.1.3 to S. If $S^{FSM} = \emptyset$ then the FSM language equation $A \diamond X \subseteq C \cup \overline{(IO)^\star}$ has no solution.*

Proof. The first step of Procedure 3.1.3 computes the intersection of S with $((I_2 \cup U)(V \cup O_2))^\star$ to enforce that the solution, if it exists, is an FSM language with input alphabet $I_2 \cup U$ and output alphabet $V \cup O_2$. Since A and C are regular languages, $S \cap ((I_2 \cup U)(V \cup O_2))^\star$ is a regular language too and, by construction, Procedure 3.1.3 extracts the largest FSM language contained in it. □

Example 3.28. Consider the FSMs $M_A = \langle S_A, I_1 \cup V, U \cup O_1, T_A, sa\rangle$ and $M_C = \langle S_C, U, V, T_C, s1\rangle$ with $S_A = \{sa\}$, $T_A = \{(i,\ sa,\ sa,\ o_2), (v,\ sa,\ sa,\ u)\}$, $S_C = \{s1\}$, $T_C = \{(i,\ s1,\ s1,\ o_1)\}$. The equation $M_A \diamond M_X \preceq M_C$ yields the language equation $A \diamond X \subseteq C \cup \overline{(IO)^\star}$ whose solution S becomes empty under prefix-closure, because S does not contain ϵ, even though it contains the string uv. Thus there is no solution.

By Proposition 3.7, it is easy to derive an FSM $M_{S^{FSM}}$ associated to S^{FSM}. This allows us to talk about FSMs that are solutions of FSM equations, meaning any reduction of the FSM $M_{S^{FSM}}$, as guaranteed by Proposition 3.8.

Example 3.29. Consider the equation $M_A \diamond M_X \preceq M_C$, with the language of M_A, i.e., $A = L_r^{\cup}(M_A)$, and the language of M_C, i.e., $C = L_r^{\cup}(M_C)$, represented by the

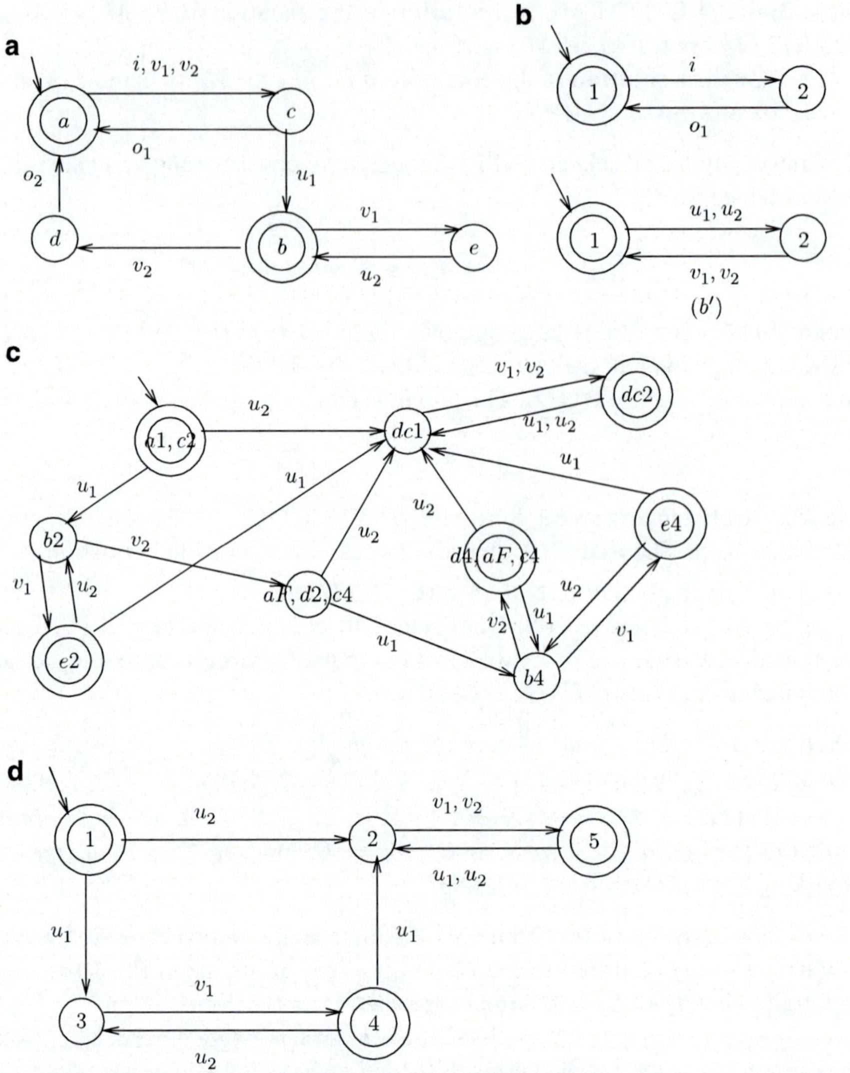

Fig. 3.10 Illustration of Example 3.29. **(a)** FA of $A = L_r^{\cup}(M_A)$; **(b)** FA of $C = L_r^{\cup}(M_C)$; **(b′)** FA of $(UV)^{\star}$; **(c)** FA of $(A \cap (\overline{C} \cap (IO)^{\star})_{\Uparrow U \cup V})_{\Downarrow U \cup V} \cap (UV)^{\star}$; **(d)** FA of prefix-closure of $(A \cap (\overline{C} \cap (IO)^{\star})_{\Uparrow U \cup V})_{\Downarrow U \cup V} \cap (UV)^{\star}$

automata shown, respectively, in Fig. 3.10a,b. The automata generating the largest language solution, $(A \cap (\overline{C} \cap (IO)^{\star})_{\Uparrow U \cup V})_{\Downarrow U \cup V} \cap (UV)^{\star}$ and its largest prefix closure are portrayed, respectively, in Fig. 3.10c,d.

Figs. 3.11 and 3.12 show the intermediate steps of the computation. Notice that in Fig. 3.10c there are two don't care states, *dc1* non-accepting and *dc2* accepting, obtained by "splitting" the accepting *dc* state of the automaton recognizing the language $(A \cap (\overline{C} \cap (IO)^{\star})_{\Uparrow U \cup V})_{\Downarrow U \cup V}$ in Fig. 3.12b, when it is intersected with

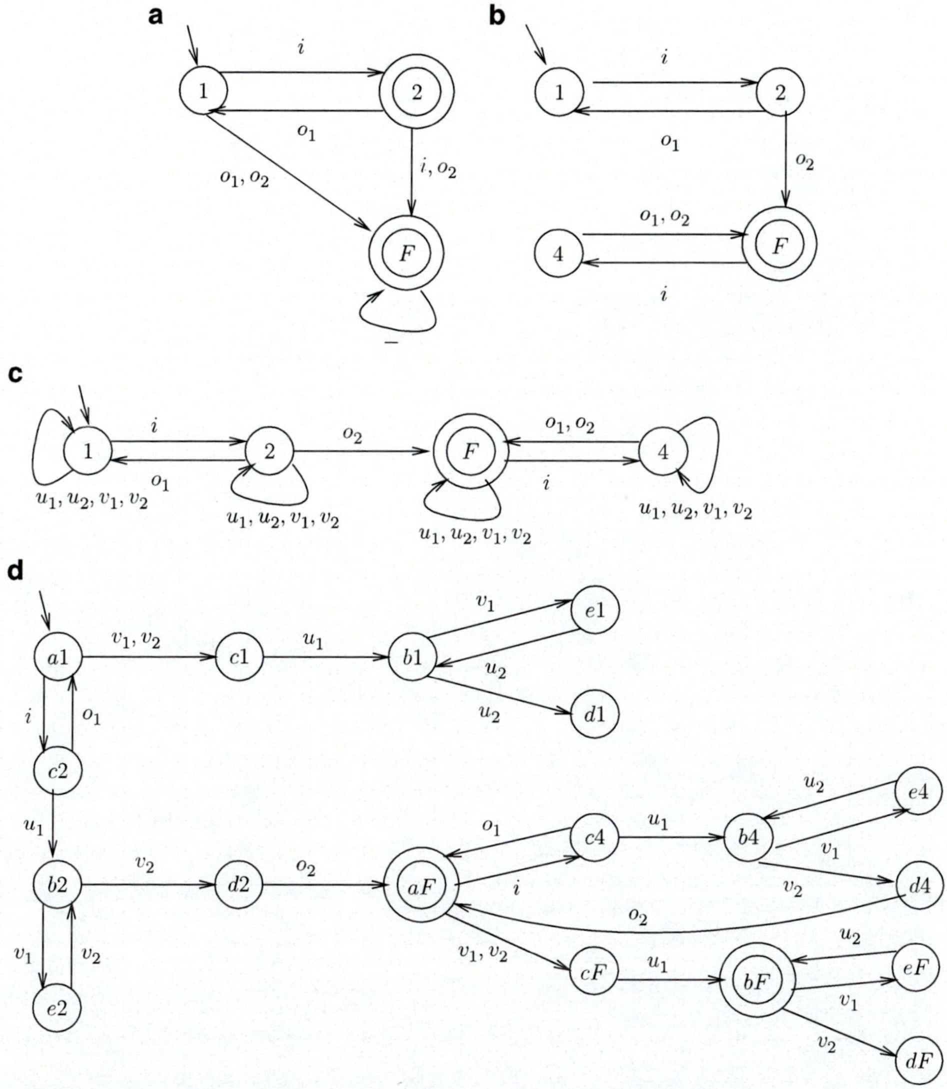

Fig. 3.11 Illustration of Example 3.29. (**a**) FA of $\overline{C}$; (**b**) FA of $\overline{C} \cap (IO)^{\star}$; (**c**) FA of $(\overline{C} \cap (IO)^{\star})_{\Uparrow U \cup V}$; (**d**) FA of $A \cap (\overline{C} \cap (IO)^{\star})_{\Uparrow U \cup V}$

the automaton of $(UV)^{\star}$ in Fig. 3.10b′. The *dc* state of the former automaton is split into states *dc1* and *dc2* because $(UV)^{\star}$ restricts the acceptance to words that terminate by a symbol in V; therefore the original *dc* state that can accept any word in $(U \cup V)^{\star}$ is replaced by two states to accept only the words in $(UV)^{\star} \subseteq (U \cup V)^{\star}$.

For logic synthesis applications, we assume that M_A and M_C are complete FSMs and we require that the solution is a complete FSM too. This is obtained by applying Procedure 3.1.4 to S^{FSM}, yielding $Prog(S^{FSM})$, the largest $(I_2 \cup U)(V \cup O_2)$-progressive FSM language $\subseteq ((I_2 \cup U)(V \cup O_2))^{\star}$.

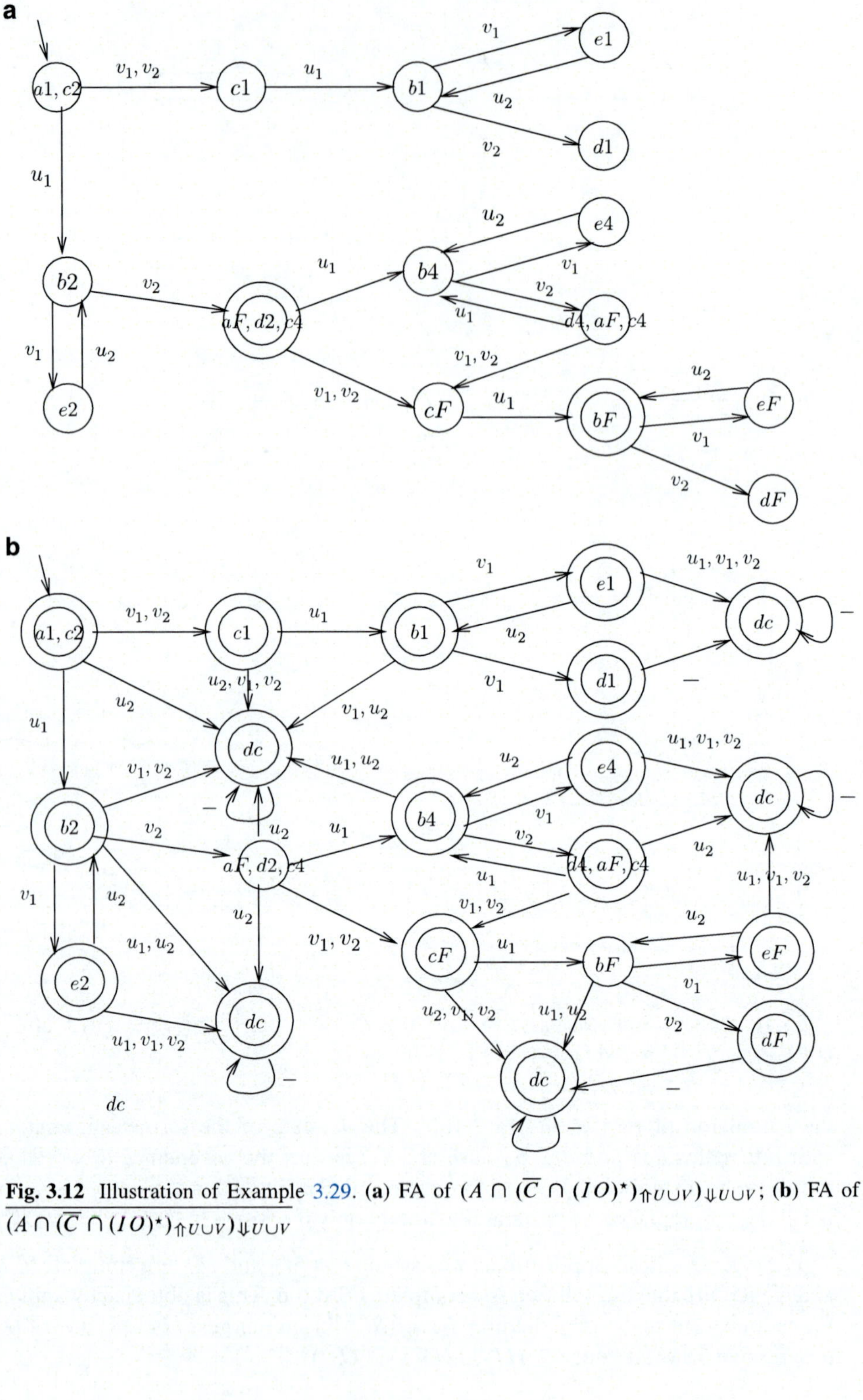

Fig. 3.12 Illustration of Example 3.29. **(a)** FA of $(A \cap (\overline{C} \cap (IO)^{\star})_{\Uparrow U \cup V})_{\Downarrow U \cup V}$; **(b)** FA of $\overline{(A \cap (\overline{C} \cap (IO)^{\star})_{\Uparrow U \cup V})_{\Downarrow U \cup V}}$

Proposition 3.30. *FSM M_B is a solution of the equation $M_A \diamond M_X \preceq M_C$, where M_A and M_C are FSMs, iff M_B is a reduction of the FSM $M_{S^{FSM}}$ associated to S^{FSM}, where S^{FSM} is obtained by applying Procedure 3.1.3 to S, where $S = \overline{A \diamond (\overline{C} \cap (IO)^{\star})}$. If $S^{FSM} = \emptyset$ then no FSM solution exists.*

The largest complete FSM solution $M_{Prog(S^{FSM})}$ is found, if it exists, by Procedure 3.1.4.

A complete FSM is a solution of $M_A \diamond M_X \preceq M_C$ iff it is a reduction of the largest complete solution $M_{Prog(S^{FSM})}$.

Given the largest (complete) solution $M_{Prog(S^{FSM})}$ of the equation $M_A \diamond M_X \preceq M_C$, let $M_A \diamond M_X \cong M_C$. Then $M_{Prog(S^{FSM})}$ is the largest (complete) solution of the equation $M_A \diamond M_X \cong M_C$.

The worst-case complexity of computing the largest solution of a parallel equation is of $2^{|S_A|.2^{|S_C|}}$, the same as for a synchronous equation. The same analysis applies (restriction plays the same role as projection in introducing nondeterminism).

3.4.2 Restricted FSM Compositional Solutions

It is interesting to compute the subset of compositionally $I^{\star}O$-progressive solutions B, i.e., such that $A_{\Uparrow I_2 \cup O_2} \cap B_{\Uparrow I_1 \cup O_1} \cap (IO)^{\star}_{\Uparrow U \cup V}$ is an $I(U \cup V)^{\star}O$-progressive FSM language $\subseteq (I(U \cup V)^{\star}O)^{\star}$. Thus the composition (after restriction to $I \cup O$) is the language of a complete FSM over inputs $I_1 \cup I_2$ and outputs $O_1 \cup O_2$. Since $A_{\Uparrow I_2 \cup O_2} \cap B_{\Uparrow I_1 \cup O_1} \cap (IO)^{\star}_{\Uparrow U \cup V}$ (after restriction to $I \cup O$) is IO-prefix-closed and hence corresponds to a partial FSM, we have to restrict it so that it is also $I(U \cup V)^{\star}O$-progressive, which corresponds to a complete FSM.

If S^{FSM} is compositionally $I(U \cup V)^{\star}O$-progressive, then S^{FSM} is the largest compositionally $I(U \cup V)^{\star}O$-progressive solution of the equation. However, not every non-empty subset of S^{FSM} inherits the feature of being compositionally $I(U \cup V)^{\star}O$-progressive. If S^{FSM} is not compositionally $I(U \cup V)^{\star}O$-progressive, then denote the largest compositionally $I(U \cup V)^{\star}O$-progressive subset of S^{FSM} by $cI(U \cup V)^{\star}OProg(S^{FSM})$. Conceptually, the language $cI(U \cup V)^{\star}OProg(S^{FSM})$ is obtained from S^{FSM} by deleting each string α such that, for some $i \in I$, there is no $(u \cup v)^{\star} \in (U \cup V)^{\star}$ and no $o \in O$ for which $\alpha\, i(u \cup v)^{\star}o \in A_{\Uparrow I_2 \cup O_2} \cap S^{FSM}_{\Uparrow I_1 \cup O_1} \cap (IO)^{\star}_{\Uparrow U \cup V}$ holds. We expect that a procedure to compute the largest compositionally $I(U \cup V)^{\star}O$-progressive prefix-closed solution, $cProg(S^{FSM})$, can be designed following the pattern of Procedure 3.3.1, but as yet have not worked out the details. A procedure to compute the largest compositionally progressive solution of a parallel equation over regular languages for the rectification topology was provided in [78].

To characterize subsets of solutions well-behaved with respect to deadlocks and livelocks (endless cycles of internal actions), we introduce a few more language definitions.

Definition 3.4.2. A solution B of (3.8) is **A-compositionally prefix $I(U \cup V)^\star O$-progressive** if

$$Pref(A)_{\Uparrow I_2 \cup O_2} \cap Pref(B)_{\Uparrow I_1 \cup O_1} \cap Pref((IO)^\star)_{\Uparrow U \cup V}$$

is $I(U \cup V)^\star O$-progressive.

A compositionally prefix $I(U \cup V)^\star O$-progressive solution yields a composition that allows $(u \cup v)^\star$ cycles without exit, yet every sequence in $I(U \cup V)^\star O$ followed by an input in I must be followed by a $(u \cup v)^\star$ cycle that can be exited (by an output).

Definition 3.4.3. A solution B of (3.8) is **A-compositionally prefix $(U \cup V)$-deadlock-free** if

$$Pref(A)_{\Uparrow I_2 \cup O_2} \cap Pref(B)_{\Uparrow I_1 \cup O_1} \cap Pref((IO)^\star)_{\Uparrow U \cup V}$$

is $(U \cup V)$-deadlock-free.

A compositionally prefix $(U \cup V)$-deadlock-free solution yields a composition that has no $(u \cup v)^\star$ cycles without exit.

Definition 3.4.4. A solution B of (3.8) is **A-compositionally prefix $(U \cup V)$-convergent** if

$$Pref(A)_{\Uparrow I_2 \cup O_2} \cap Pref(B)_{\Uparrow I_1 \cup O_1} \cap Pref((IO)^\star)_{\Uparrow U \cup V}$$

is $(U \cup V)$-convergent.

A compositionally prefix $(U \cup V)$-convergent solution yields a composition that has no $(u \cup v)^\star$ cycles, i.e., it is livelock-free. A compositionally prefix $(U \cup V)$-deadlock-free solution does not need to be compositionally prefix $(U \cup V)$-convergent.

Example 3.31. Consider the equation $M_A \diamond M_X \preceq M_C$, where FSMs M_A and M_C and the largest solution M_B are shown in Fig. 3.13a–c. Fig. 3.13d,e shows the related automata A and $B_{\Uparrow I_1 \cup O_1} = B_{\Uparrow \{i\} \cup \{o\}}$, whereas Fig. 3.13f,g portrays the automata representing the languages $A_{\Uparrow I_2 \cup O_2} \cap B_{\Uparrow I_1 \cup O_1} \cap (IO)^\star_{\Uparrow U \cup V} = A \cap B_{\Uparrow \{i\} \cup \{o\}} \cap (IO)^\star_{\Uparrow \{u\} \cup \{v\}}$ and $Pref(A)_{\Uparrow I_2 \cup O_2} \cap Pref(B)_{\Uparrow I_1 \cup O_1} \cap Pref((IO)^\star)_{\Uparrow U \cup V} = Pref(A) \cap Pref(B)_{\Uparrow \{i\} \cup \{o\}} \cap Pref((IO)^\star)_{\Uparrow \{u\} \cup \{v\}}$.

If FSM M_A answers by u to the external input i then FSMs M_A and M_B fall into an infinite dialog, so we would like to classify their composition as neither $(U \cup V)$-convergent nor $(U \cup V)$-deadlock-free. However the language $A \cap B_{\Uparrow \{i\} \cup \{o\}} \cap (IO)^\star_{\Uparrow \{u\} \cup \{v\}} = \{(io)^\star\}$ is both $(U \cup V)$-convergent and $(U \cup V)$-deadlock-free. To overcome this modeling problem, we introduce the operator $Pref$ (guarantees prefix-closure) and rewrite the previous language composition as $Pref(A) \cap Pref(B)_{\Uparrow \{i\} \cup \{o\}} \cap Pref((IO)^\star)_{\Uparrow \{u\} \cup \{v\}}$. The latter language is

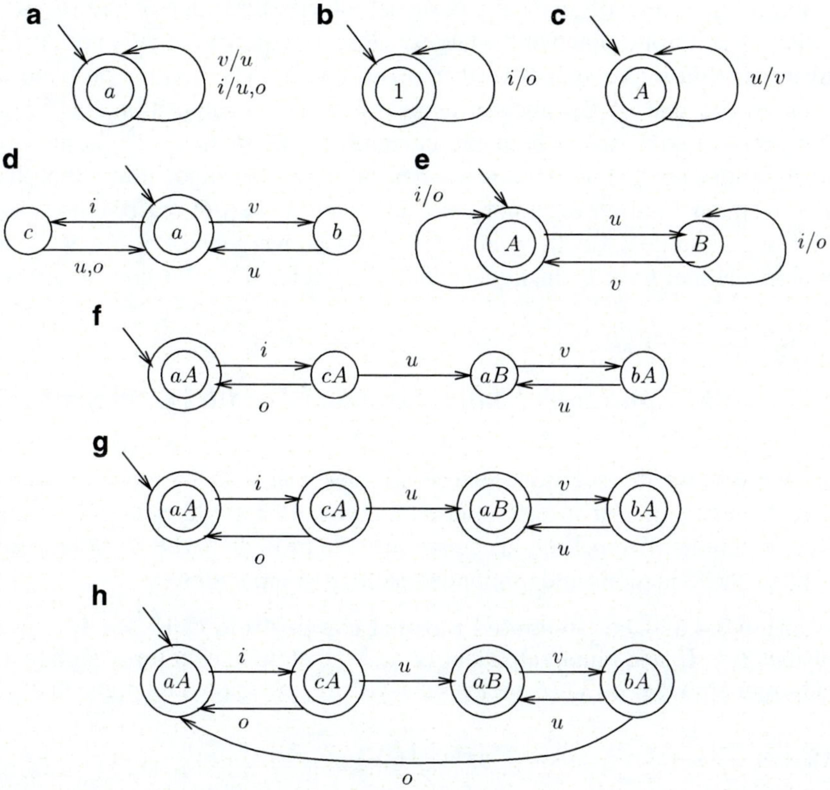

Fig. 3.13 Illustration of Example 3.31. **(a)** FSM M_A; **(b)** FSM M_C; **(c)** FSM M_B; **(d)** FA of $A = L_r^{\cup}(M_A)$; **(e)** FA of $B_{\Uparrow\{i\}\cup\{o\}}$, where $B = L_r^{\cup}(M_B)$; **(f)** FA of $A \cap B_{\Uparrow\{i\}\cup\{o\}} \cap (IO)^{\star}_{\Uparrow\{u\}\cup\{v\}}$; **(g)** FA of $Pref(A) \cap Pref(B)_{\Uparrow\{i\}\cup\{o\}} \cap Pref((IO)^{\star})_{\Uparrow\{u\}\cup\{v\}}$; **(h)** FA of prefix $(U \cup V)$-deadlock-free, but not prefix $(U \cup V)$-convergent language

neither $(U \cup V)$-convergent (since $i_{\Uparrow\{u\}\cup\{v\}}$ includes $iu(v+u)^{\star}$ that is a subset of the language) nor $(U \cup V)$-deadlock-free ($\alpha i u$ cannot be extended to a string ending by o, against the definition of $(U \cup V)$-deadlock-free).

Finally Fig. 3.13h shows a language that is $(U \cup V)$-deadlock-free, but not $(U \cup V)$-convergent.

Theorem 3.32. *Let B be an $(I_2 \cup U)(V \cup O_2)$-progressive solution of $A \diamond X \subseteq C \cup \overline{(IO)^{\star}}$ and let A be $(I_1 \cup V)(U \cup O_1)$-progressive. If B is compositionally prefix $(U \cup V)$-convergent, then B is compositionally prefix $(U \cup V)$-deadlock-free.*

Proof. Since the components A and B are progressive, their composition $Pref(A)_{\Uparrow I_2 \cup O_2} \cap Pref(B)_{\Uparrow I_1 \cup O_1} \cap Pref((IO)^{\star})_{\Uparrow U \cup V}$ is deadlock-free, i.e., it never stops because a component does not have a transition under a given input. If the composition is also $(U \cup V)$-convergent, there can be no livelocks, i.e., there are no cycles labeled with actions from the set $U \cup V$. Therefore an external input, after a finite path labeled with internal actions, must be followed by an external output. □

The computation of the largest subset of compositionally prefix $(U \cup V)$-deadlock-free solutions and of the largest subset of compositionally prefix $(U \cup V)$-convergent solutions requires further investigation. The former problem appears similar to the one of finding the largest subset of compositionally $I(U \cup V)^{\star}$ O-progressive solutions. About the latter problem, when S^{FSM} is not compositionally prefix $(U \cup V)$-convergent, then the largest compositionally prefix $(U \cup V)$-convergent solution does not exist and each finite IO-prefix-closed subset of S^{FSM} is a compositionally prefix $(U \cup V)$-convergent solution. It is an open question whether there is the largest complete prefix $(U \cup V)$-convergent solution.

3.4.3 FSM Equations Under Bounded Parallel Composition

Here we discuss the solutions whose composition with the context produces an external output after at most l internal actions. One could build an analogy with Moore solutions of synchronous equations. We provide in the sequel the key steps to solve FSM equations under bounded parallel composition.

Definition 3.4.5. The **l-bounded parallel composition of FSMs** M_B, over input alphabet $I_2 \cup U$ and output alphabet $O_2 \cup V$, with M_A, over input alphabet $I_1 \cup V$ and output alphabet $O_1 \cup U$, yields the FSM $M_A \diamond_l M_B$ with language

$$L(M_A \diamond_l M_B) = L(M_A) \diamond_l L(M_B) \cap (IO)^{\star}$$

$$= \left[L(M_A)_{\Uparrow I_2 \cup O_2} \cap L(M_B)_{\Uparrow I_1 \cup O_1} \cap (I \cup O)^{\star}_{\Uparrow (U \cup V, l)} \right]_{\Downarrow I \cup O} \cap (IO)^{\star}.$$

When $l = \infty$, it reduces to the definition of parallel composition of FSMs.

Example 3.33. Fig. 3.14a,b show the automata of $(I \cup O)^{\star}_{\Uparrow (U \cup V, l=1)}$ and $(I \cup O)^{\star}_{\Uparrow (U \cup V, l=2)}$, with $I = \{i\}$, $O = \{o\}$, $U = \{u\}$ and $V = \{v\}$. The words in $(I \cup O)^{\star}_{\Uparrow (U \cup V, 1)}$ are obtained from those in $(I \cup O)^{\star}$ by inserting anywhere in them words from $(U \cup V)^{\leq 1} = \{\epsilon, u, v\}$; the words in $(I \cup O)^{\star}_{\Uparrow (U \cup V, 2)}$ are obtained from those in $(I \cup O)^{\star}$ by inserting anywhere in them words from $(U \cup V)^{\leq 2} = \{\epsilon, u, v, uu, uv, vu, vv\}$.

Proposition 3.34. *FSM M_B is a solution of the equation $M_A \diamond_l M_X \preceq M_C$, where M_A and M_C are FSMs, iff M_B is a reduction of the FSM $M_{S^{FSM}}$ associated with S^{FSM}, where S^{FSM} is obtained by applying Procedure 3.1.3 to S, where $S = \overline{(A_{\Uparrow I_2 \cup O_2} \cap (\overline{C} \cap (IO)^{\star})_{\Uparrow (U \cup V, l)})_{\Downarrow I_2 \cup U \cup V \cup O_2}}$.*

If $S^{FSM} = \emptyset$ then no FSM is a solution. S^{FSM} is the largest compositionally $(U \cup V)$-convergent solution of $M_A \diamond_l M_X \preceq M_C$.

The largest complete FSM solution $M_{Prog(S^{FSM})}$ is found, if it exists, by Procedure 3.1.4.

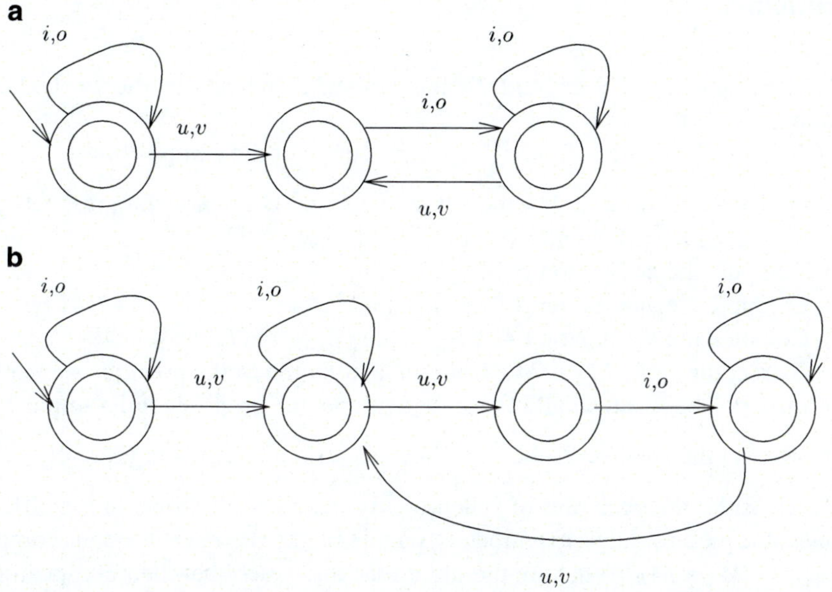

Fig. 3.14 Illustration of Example 3.33. (**a**) FA of $(I \cup O)^{\star}_{\Uparrow(U \cup V, l=1)}$; (**b**) FA of $(I \cup O)^{\star}_{\Uparrow(U \cup V, l=2)}$. In both cases it is $I = \{i\}$, $O = \{o\}$, $U = \{u\}$ and $V = \{v\}$

Theorem 3.35. *A solution M_B of $M_A \diamond_l M_X \preceq M_C$ is also a compositionally $(U \cup V)$-convergent solution of $M_A \diamond M_X \preceq M_C$.*

If M_A and M_B are also complete FSMs, then M_B is a compositionally prefix $I(U \cup V)^{\star}O$-progressive and compositionally $I^{\star}O$-progressive solution of $M_A \diamond M_X \preceq M_C$.

Proof. By construction, a solution M_B of $M_A \diamond_l M_X \preceq M_C$ is compositionally $(U \cup V)$-convergent. A solution M_B of $M_A \diamond_l M_X \preceq M_C$ is also a solution of $M_A \diamond M_X \preceq M_C$, because when $l = \infty$ the operator $\diamond_l$ becomes the operator $\diamond$.

By Theorem 3.32, the fact that M_B is compositionally $(U \cup V)$-convergent, together with the completeness of M_A and M_B, imply that M_B is compositionally prefix $I(U \cup V)^{\star}O$-progressive and therefore compositionally $I^{\star}O$-progressive. □

However, in general $M_A \diamond M_X \preceq M_C$ may be solvable despite the fact that $M_A \diamond_l M_X \preceq M_C$ has no solution. For instance, this may happen when $M_A \diamond M_X \preceq M_C$ has no compositionally $I(U \cup V)^{\star}O$-progressive solution. If the equation $M_A \diamond_l M_X \preceq M_C$ has no complete solution, it is open whether there is a compositionally $I(U \cup V)^{\star}O$-progressive solution of $M_A \diamond M_X \preceq M_C$.

Problems

3.1. Consider the FSMs and their parallel composition shown textually in Example 3.14(a).

Compute all the steps to verify the correctness of the composition.

(a) Derive from M_A and M_B the automata A and B recognizing, respectively, the languages $L_r^{\cup}(M_A)$ and $L_r^{\cup}(M_B)$.
(b) Compute the automaton $B_{\Uparrow I_1 \cup O_1}$.
(c) Compute the automaton $A \cap B_{\Uparrow I_1 \cup O_1}$.
(d) Compute the automaton $(A \cap B_{\Uparrow I_1 \cup O_1})_{\Downarrow I_1 \cup O_1} \cap (I_1 O_1)^{\star}$.
(e) Extract the FSM $M_A \diamond M_B$ from the automaton in the previous step and minimize it. The result should coincide with the one proposed in Example 3.14(a).

3.2. Repeat the steps of Problem 3.1 for the variant reported in Example 3.14(b).

3.3. Consider the examples of synchronous composition shown in Fig. 3.1. For instance, the picture shows machines M_A and M_B and their synchronous composition $M_{A \bullet B} = M_A \bullet M_B$. What are the other machines such that their composition with the context M_A yields the same composed machine $M_{A \bullet B}$?
Compute all such machines by solving the equation $M_A \bullet M_X = M_{A \bullet B}$. What is the relation between the largest solution of the equation and the given machine M_B?

3.4. Consider the traffic controller example described in [12] and already introduced in Problem 2.1.

Interpret the state graphs in Figs. 2.10 and 2.11 as respectively the context FSM M_A and the specification FSM M_C, by defining their input and output variables as follows. The inputs of M_A are two binary variables v_1, v_2, and its output is the multi-valued variable *colours* that can assume the values *green, red, yellow*. The inputs of M_C are two multi-valued variables i_1, i_2 that can assume each one of the three values $1, 2, 3$, its output is again the multi-valued variable *colours*.

Find the largest solution of the equation $M_A \bullet M_X \subseteq M_C$, where the input variables of M_X are i_1, i_2 and its output variables are v_1, v_2. This is a series topology where M_X feeds M_A, and the composition of M_X and M_A yields M_C.

Repeat the problem assuming that M_X has one more input: the variable *colours* produced by M_A, as in the controller's topology.

3.5. Consider the FSMs $M_A = \langle S_A, V, U, T_A, sa \rangle$ and $M_C = \langle S_C, I, O, T_C, s1 \rangle$ with $S_A = \{sa, sb\}$, $T_A = \{(v1,\ sa,\ sb,\ u1), (v2.\ sa,\ sa,\ u1), (v1,\ sb,\ sa,\ u1), (v2,\ sb,\ sb,\ u2)$, and $S_C = \{s1, s2, s3\}$, $T_C = \{(i1,\ s1,\ s2,\ o1), (i2,\ s1,\ s1,\ o2), (i1,\ s2,\ s3,\ o2), (i2,\ s2,\ s3,\ o1)\}, (i2,\ s3,\ s3,\ o1), (i1,\ s3,\ s1,\ o2)$.

Compute the largest solution of the equation $M_A \bullet M_X \subseteq M_C$, where M_X has inputs $I \times U$ and outputs $V \times O$. Notice that in this example only M_X has access to the environment.

3.6. Consider the FSMs $M_A = \langle S_A, V, O, T_A, sa \rangle$ and $M_C = \langle S_C, I, O, T_C, s1 \rangle$ with $S_A = \{sa, sb\}$, $T_A = \{(v1,\ sa,\ sb,\ o1), (v2.\ sa,\ sb,\ o2), (v1,\ sb,\ sa,\ o2),$

$(v2, sb, sb, o2)$, and $S_C = \{s1, s2, s3, s4\}$, $T_C = \{(i2, s1, s2, o1), (i1, s1, s3, o1),$ $(i1, s1, s3, o2), (i1, s2, s3, o2)\}, (i2, s2, s4, o2), (i2, s3, s3, o2)\},$ $(i1, s3, s1, o2), (i1, s4, s2, o1)\}, (i2, s4, s3\, o2)$.

(a) Compute the largest solution of the equation $M_A \bullet M_X \subseteq M_C$, where M_X has inputs $I \times O$ and outputs V. This is an example of controller's topology.
(b) Compute the largest solution of the equation $M_A \bullet M_X \subseteq M_C$, where M_X has inputs I and outputs V. This is an example of series topology.
(c) Compare the solutions obtained in (b) and (c).

Chapter 4
Equations Over ω-Automata

4.1 ω-Languages and ω-Automata

An infinite word over an alphabet A, or ω-word, is an infinite sequence of symbols of A. A^ω is the set of ω-words on A. An ω-language on A is a subset of A^ω. Moreover, $A^\infty = A^\star \cup A^\omega$. An ω-word may be written as $\alpha = \alpha(0)\alpha(1)\ldots$, where $\alpha(i) \in A$ for every $i \geq 0$; if $n \leq m$, $\alpha(n,m) = \alpha(n)\ldots\alpha(m-1)\alpha(m)$ and $\alpha(n,\infty) = \alpha(n)\alpha(n+1)\ldots$. The notations $\exists^\omega n$ stands for 'there are infinitely many n' and $\exists^{<\omega} n$ stands for 'there are finitely many n'. Given a set $W \subseteq A^\star$, we define the sets $W^\omega = \{\alpha \in A^\omega \mid \alpha = w_0 w_1 \ldots,\ w_i \in W\}$, $\overrightarrow{W} = \{\alpha \in A^\omega \mid \exists^\omega n\ \alpha(0,n) \in W\}$ (also called $lim(W)$), and $Inf(\alpha) = \{a \in A \mid \exists^\omega n\ \alpha(n) = a\}$. Given an ω-language L, $Pref(L)$ denotes (by overloading) the sets of finite prefixes of the words in L.

Definition 4.1.1. A **ω-automaton** is a 5-tuple $\mathcal{A} = (Q, A, \Delta, q_0, \phi)$, where ϕ is a Boolean formula defining the acceptance conditions.[1]

A **run** of $\mathcal{A}$ under an ω-word α is an ω-word r_α over the alphabet of the states in Q such that $r_\alpha(0) = q_0$, and, for every $i \geq 0$, $(r_\alpha(i), \alpha(i), r_\alpha(i+1)) \in \Delta$.

A run r_α under an ω-word α is **accepting** if $\phi(\mathrm{Inf}(r_\alpha))$ is true, i.e., the set of states of r_α visited infinitely often satisfies the condition ϕ.

An ω-word α is **accepted** if there exists an accepting run r_α under α.

The language **accepted** by the ω-automaton $\mathcal{A}$ with acceptance condition ϕ, $L(\mathcal{A})$, is the set of accepted strings $\alpha \in A^\omega$, i.e., the strings for which there is a run of states r_α visited by α such that $\phi(\mathrm{Inf}(r_\alpha))$ is true (the set of states visited infinitely often by r_α satisfies the condition ϕ).

[1] For a Büchi automaton ϕ is a disjunctive formula representing the subset of states $F \subseteq Q$; for a co-Büchi automaton ϕ is $\neg\overline{F}$, where $\overline{F}$ is disjunctive formula representing the set $Q \setminus F$; for a Muller automaton ϕ is $\vee_{F\in\mathcal{F}}(\wedge_{f\in F} f\ \ \wedge_{q\notin F} \neg q)$, where $\mathcal{F} \subseteq 2^Q$; for a Rabin automaton ϕ is $\vee_{i=1}^n(L_i \wedge \neg(\overline{U_i}))$, where $L_i, U_i, 1 \leq i \leq n$, are disjunctive formulas; for a Street automaton ϕ is $\wedge_{i=1}^n(L_i \wedge \neg(\overline{U_i}))$, where $L_i, U_i, 1 \leq i \leq n$, are disjunctive formulas.

T. Villa et al., *The Unknown Component Problem: Theory and Applications*,
DOI 10.1007/978-0-387-68759-9_4,

Definition 4.1.2. A **Büchi automaton** is a 5-tuple $\mathcal{A} = (Q, A, \Delta, q_0, F)$, where Q is a finite set of states, A is a finite set of symbols, $q_0 \in Q$ is the initial state, $F \subseteq Q$ is the set of final states, and $\Delta \subseteq Q \times A \times Q$ is the transition relation.

A run σ over an ω-word α is **accepting** if $\mathrm{Inf}(\sigma) \cap F \neq \emptyset$. A language is **$\omega$-regular** if and only if it is accepted by a Büchi automaton.

The final states of a finite automaton can be interpreted as the recurrent states of a Büchi automaton, where accepting runs visit final states infinitely often. We get a *co-Büchi automaton* if the acceptance condition identifies a subset of states called *co-final states* or *persistent states*, where an accepting run finally never leaves the co-final states. The final states of a finite automaton can be interpreted as the persistent states of a co-Büchi automaton, where accepting runs eventually stay in the set of final states.

The following closure properties hold for ω-languages.

Proposition 4.1. *1. If $V \subseteq A^\star$ is regular, then V^ω is ω-regular.*
2. If $V \subseteq A^\star$ is regular and $L \subseteq A^\omega$ is ω-regular, then $V.L$ is ω-regular.
3. If $L_1, L_2 \subseteq A^\omega$ are ω-regular, then $L_1 \cup L_2$ and $L_1 \cap L_2$ are ω-regular.

Moreover, there are effective constructions to obtain the automata of the resulting languages.

Definition 4.1.3. An **ω-regular expression** has the form $\bigcup U_i.V_i^\omega$, where for all $1 \leq i \leq n$, U_i and V_i are regular expressions.

Theorem 4.2 (Büchi - 1960). *For every Büchi automaton there is an ω-regular expression that defines the same ω-language and vice versa.*

Definition 4.1.4. A word $\alpha \in A^\omega$ is **definitively periodic** if there are $u, v \in A^\star$ such that $\alpha = uv^\omega$.

Corollary 4.3. *Every non-empty ω-regular language contains an infinitely periodic ω-word.*

The language of a Büchi automaton is non-empty if and only if its transition graph has a cycle that contains a final state reachable from the initial state.

Theorem 4.4 (Büchi - 1960). *If $L \subseteq A^\omega$ is ω-regular, then $\overline{L} = A^\omega \setminus L$ is ω-regular. There is an effective construction to build a Büchi automaton accepting $\overline{L}$ from the Büchi automaton accepting L.*

However the construction is computationally very hard, as discussed in [42,53,122].

Büchi automata over infinite words might appear as a direct extension of finite automata from finite words to infinite words, sharing the same underlying transition structure with acceptance conditions made infinitary instead than finitary. However the following examples highlight the fundamental differences.

Example 4.5. Figure 4.1 shows three examples of automata $\mathcal{A}_1$, $\mathcal{A}_2$ and $\mathcal{A}_3$, with the same ω-language $\mathcal{L}^\omega(\mathcal{A}_1) = \mathcal{L}^\omega(\mathcal{A}_2) = \mathcal{L}^\omega(\mathcal{A}_3) = \{(ab)^\omega\}$; however, their respective finitary languages are $\mathcal{L}^{\mathrm{fin}}(\mathcal{A}_1) = \{(ab)^\star(a + \epsilon)\}$, $\mathcal{L}^{\mathrm{fin}}(\mathcal{A}_2) = \{(ab)^\star\}$,

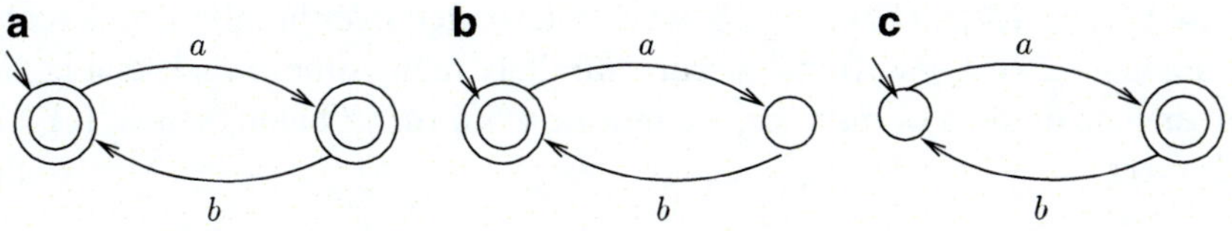

Fig. 4.1 (**a**) Automaton $\mathcal{A}_1$; (**b**) Automaton $\mathcal{A}_2$; Automaton $\mathcal{A}_3$

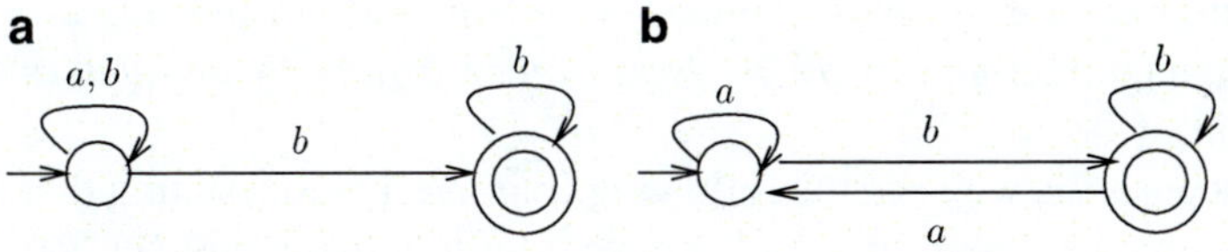

Fig. 4.2 (**a**) Non-deterministic automaton $\mathcal{A}_1$; (**b**) deterministic automaton $\mathcal{A}_2$

$\mathcal{L}^{\text{fin}}(\mathcal{A}_3) = \{a(ba)^\star\}$. Notice that ω-regular languages, as $\star$-regular languages, can be denoted by different expressions, e.g., $\mathcal{L}^\omega(\mathcal{A}_3) = \{(ab)^\omega\} = \{a(ba)^\omega\}$.[2]

Example 4.6. The equality of prefixes of ω-languages does not imply the equality of ω-languages, nor vice versa, as the following examples show.[3]

- From $L_1^\omega = L_2^\omega$, it does not follow $\mathit{Pref}(L_1^\omega) = \mathit{Pref}(L_2^\omega)$, i.e., the equality of ω-languages does not imply the equality of prefixes. E.g., given two empty ω-languages L_1^ω and L_2^ω, it may be that $\mathit{Pref}(L_1^\omega) = \{ab\}$, $\mathit{Pref}(L_2^\omega) = \{ba\}$ and so $\mathit{Pref}(L_1^\omega) \neq \mathit{Pref}(L_2^\omega)$. Question: does this happen only when $L_1^\omega = L_2^\omega = \emptyset$?
- From $\mathit{Pref}(L_1^\omega) = \mathit{Pref}(L_2^\omega)$, it does not follow $L_1^\omega = L_2^\omega$, i.e., the equality of prefixes does not imply the equality of ω-languages. E.g., $\Sigma = \{a, b\}$, $L_1^\omega = \Sigma^\star a \Sigma^\omega$, $L_2^\omega = \Sigma^\star a \Sigma^\omega \cup b^\omega$, $\mathit{Pref}(L_1^\omega) = \mathit{Pref}(L_2^\omega)$, but $L_1^\omega \neq L_2^\omega$.

  Another example, is $\mathit{Pref}(L_1^\omega) = \mathit{Pref}(L_2^\omega) = a^\star b^\star$, and $L_1^\omega = a^\star b^\omega$, $L_2^\omega = a^\omega + a^\star b^\omega$.

Example 4.7. Consider the *non-deterministic* automaton $\mathcal{A}_1$ in Fig. 4.2a.[4]
The language of $\mathcal{A}_1$ when interpreted as an automaton over finite words is $\mathcal{L}^{\text{fin}}(\mathcal{A}_1) = \{w : w \text{ ends with } b\}$, whereas its language when interpreted over infinite words is $\mathcal{L}^\omega(\mathcal{A}_1) = \{w : a \text{ occurs a finite number of times in } w\}$.

For languages over finite words, we know that non-deterministic automata and deterministic automata are equivalent. Hence, consider the following *deterministic* automaton $\mathcal{A}_2$ in Fig. 4.2b.

$\mathcal{A}_2$ is obtained by determinization of the automaton $\mathcal{A}_1$, by applying the classic subset construction introduced for automata over finite words. Therefore $\mathcal{A}_2$ when interpreted as an automaton over finite words recognizes the same language as $\mathcal{A}_1$:

---

[2]We thank V. Bushkov, University of Tomsk, for discussions on Example 4.5.

[3]We thank A. Chebotarev, Ukrainian Academy of Sciences, Kiev, for discussions on Example 4.6.

[4]We thank D. Bresolin, University of Verona, for discussions on Example 4.7.

$\mathcal{L}^{\text{fin}}(\mathcal{A}_2) = \mathcal{L}^{\text{fin}}(\mathcal{A}_1)$. However, $\mathcal{A}_2$ when interpreted over infinite words recognizes the language $\mathcal{L}^{\omega}(\mathcal{A}_2) = \{w : b \text{ occurs infinitely many times in } w\}$, that is different from the one of $\mathcal{A}_1$. For instance, the infinite word $(ab)^{\omega}$ belongs to $\mathcal{L}^{\omega}(\mathcal{A}_2)$ but not to $\mathcal{L}^{\omega}(\mathcal{A}_1)$.

This example shows that the determinization operation for finite automata does not preserve the language recognized by a Büchi automaton. Even more, this fact can be seen as the corollary of a more general result, stating that deterministic Büchi automata are not closed under complementation. Indeed, it can be proved the ω-language recognized by $\mathcal{A}_1$ *is not recognizable by any deterministic Büchi automaton*.

To that purpose, we need the following characterization, stating that deterministic Büchi automata accept ω-words that have infinitely many prefixes accepted as finite words, so roughly speaking deterministic Büchi automata are the natural extension of automata over finite words.

Theorem 4.8. *A language $L \subseteq A^{\omega}$ is accepted by a deterministic Büchi automaton if and only if there is a regular set $W \subseteq A^{\star}$ such that $L = \overrightarrow{W}$.*

Proof (Adapted from [98]). Let $\mathcal{A} = (Q, A, \Delta, q_0, F)$ be a deterministic Büchi automaton and $\mathcal{A}' = (Q, A, \Delta, q_0, F)$ be the corresponding deterministic finite automaton ($\mathcal{A}$ and $\mathcal{A}'$ have the same underlying structure and differ only by the acceptance condition). Let $W = L(\mathcal{A}') \subseteq A^{\star}$ be the language recognized by $\mathcal{A}'$.

The thesis is that $L(\mathcal{A}) = \overrightarrow{W}$.

We prove first that $L(\mathcal{A}) \subseteq \overrightarrow{W}$. Let $\alpha \in L(\mathcal{A})$. From the acceptance condition of deterministic Büchi automata we have that $\mathcal{A}$ accepts an ω-word α if and only if the run r_{α} of $\mathcal{A}$ under α satisfies $\text{Inf}(r_{\alpha}) \cap F \neq \emptyset$, i.e., the set of states visited infinitely often by r_{α} intersects the set of final states F. It follows that there are infinitely many prefixes w of α such that $w \in W$, i.e., $\alpha \in \overrightarrow{W}$ by definition of $\overrightarrow{W}$.

Now we prove that $\overrightarrow{W} \subseteq L(\mathcal{A})$. Suppose that $\alpha \in \overrightarrow{W}$. For every prefix w of α with $w \in W = L(\mathcal{A}')$, the accepting run of $\mathcal{A}'$ over w may be seen as a prefix of the unique run r_{α} of $\mathcal{A}$ over α (unique because $\mathcal{A}$ is a deterministic Büchi automaton). Therefore $\text{Inf}(r_{\alpha}) \cap F \neq \emptyset$ and so $\alpha \in L(\mathcal{A})$. □

Notice that in the proof of Theorem 4.8 the hypothesis that the Büchi automaton is deterministic is used in proving that $\overrightarrow{W} \subseteq L(\mathcal{A})$ (whereas it is not needed in proving that $L(\mathcal{A}) \subseteq \overrightarrow{W}$).

As a check, suppose to remove that hypothesis. Then the statement that $\mathcal{A}$ accepts α if there are infinitely many prefixes of α that drive $\mathcal{A}'$ to a final state is false. Indeed from Example 4.7 consider the non-deterministic automaton $\mathcal{A}_1$ (set in this proof $\mathcal{A} = \mathcal{A}_1$) and the ω-word $\alpha = (ab)^{\omega}$; there are infinitely many prefixes of α of the form $(ab)^n, n \geq 1$, that drive $\mathcal{A}'$ to a final state and so they are finite words in W, but it is false that $\mathcal{A}$ accepts α, because there is no final state that occurs infinitely often in any run r_{α} of $\mathcal{A}$ and so there is no accepting run r_{α}.

Now we show that there is no deterministic Büchi automaton that accepts the language $\mathcal{L}^\omega(\mathcal{A}_1) = \{\alpha \in A^\omega : a \text{ occurs a finite number of times in } \alpha\} = \{\alpha \in A^\omega : \exists^{<\omega} n\ \alpha(n) = a\}$. It is enough to show that there is no regular set $W \subseteq A^\star$ such that $\mathcal{L}^\omega(\mathcal{A}_1) = \overrightarrow{W}$. Suppose by contradiction that there is such a set W. Since $b^\omega \in \mathcal{L}^\omega(\mathcal{A}_1) = \overrightarrow{W}$, there is n_1 such that $b^{n_1} \in W$ (b^ω has infinitely many prefixes in W, in particular there will be a prefix $b^{n_1} \in W$). Similarly, since $b^{n_1}ab^\omega \in \mathcal{L}^\omega(\mathcal{A}_1) = \overrightarrow{W}$, there is n_2 such that $b^{n_1}ab^{n_2} \in W$. Iterating the reasoning, we construct an ω-word $\alpha = b^{n_1}ab^{n_2}ab^{n_3}\ldots$ with infinitely many prefixes in W. By definition of $\overrightarrow{W}$, $\alpha \in \overrightarrow{W}$, but $\overrightarrow{W} = \mathcal{L}^\omega(\mathcal{A}_1)$ and so $\alpha \in \mathcal{L}^\omega(\mathcal{A}_1)$, which is a contradiction since α contains infinitely many occurrences of a and so cannot be in $\mathcal{L}^\omega(\mathcal{A}_1)$.

Hence, the following theorem holds.

Theorem 4.9. *The class of ω-languages recognized by deterministic Büchi automata is* strictly included *in the class of ω-languages recognized by non-deterministic Büchi automata.*

Moreover, deterministic Büchi automata are not closed under complementation.

Example 4.10. Consider the example in Fig. 4.2b. If we interpret it as a co-Büchi automaton with the final state interpreted as the persistent state, the language accepted is the set of ω-words with finite occurrences of a. Compare it with the discussion in Example 4.7 to understand the duality between Büchi and co-Büchi conditions.

Definition 4.1.5. A **Muller automaton** is a 5-tuple $\mathcal{A} = (Q, A, \Delta, q_0, \mathcal{F})$, where Q is a finite set of states, A is a finite set of symbols, $q_0 \in Q$ is the initial state, $\mathcal{F} \subseteq 2^Q$ is a collection of sets of final states, and $\Delta \subseteq Q \times A \times Q$ is the transition relation.

A run σ over an ω-word α is **accepting** if $\mathrm{Inf}(\sigma) \in \mathcal{F}$. The automaton $\mathcal{A}$ accepts an ω-word α if there is an accepting run of $\mathcal{A}$ over α. The language $\mathcal{L}(\mathcal{A})$ accepted by $\mathcal{A}$ is the set of ω-words accepted by $\mathcal{A}$.

Given a deterministic Büchi automaton, a deterministic Muller automaton equivalent to it can be defined with the accepting condition $\mathcal{F} = \{S \in 2^Q \mid S \cap F \neq \emptyset\}$.

We state a few basic theorems about Muller automata.

Theorem 4.11. *Non-deterministic Muller automata accept ω-regular languages and vice versa.*

Finally, the languages accepted by deterministic Muller automata are the same as the languages accepted by non-deterministic Büchi automata. So deterministic Muller automata have the same expressive power as nondeterministic Muller automata.

Theorem 4.12 (McNaughton). *An ω-language is ω-regular (i.e., accepted by a Büchi automaton) if and only if it is accepted by a deterministic Muller automaton.*

4.2 Equations Over ω-Languages

We saw that Büchi automata have an underlying structure as finite automata, augmented with an infinitary acceptance condition. Büchi automata accept ω-regular languages; two automata are ω-equivalent if they accept the same language, and automaton $\mathcal{A}$ is an ω-reduction of automaton $\mathcal{B}$ if the ω-language of $\mathcal{A}$ is a subset of the ω-language of $\mathcal{B}$.

One might conjecture that the solution to an equation over ω-languages could be obtained by solving the corresponding equation over their finitary counterparts, and then computing the limit of the latter solution to obtain the solution of the original equation over ω-languages, however the following example shows that it is not true in general.

Example 4.13. Given $I = \{i_1, i_2\}$, $U = \{u_1, u_2\}$, consider the ω-regular languages $C^\omega = ((i_1 + i_2)(u_1 + u_2))^\star(i_1u_1)^\omega + (i_2u_2)^\omega$ and $S^\omega = (i_1 + i_2)^\star i_1^\omega$.

Then the regular languages of the finite prefixes of the ω-languages C^ω and S^ω are, respectively, $Pref(C^\omega) = ((i_1 + i_2)(u_1 + u_2))^\star(i_1u_1)^\star + (i_2u_2)^\star$ and $Pref(S^\omega) = (i_1 + i_2)^\star i_1^\star = (i_1 + i_2)^\star$.

The largest solution of the equation $Pref(C^\omega) \diamond X \subseteq Pref(S^\omega)$ is the language $L = (u_1 + u_2)^\star$, whose limit is $\overrightarrow{L} = (u_1 + u_2)^\omega$.

Indeed, $L_{\Uparrow I} = ((u_1 + u_2)^\star)_{\Uparrow I} = (u_1 + u_2 + i_1 + i_2)^\star$, whose intersection with $Pref(C^\omega)$ is equal to $Pref(C^\omega)$. Then $(Pref(C^\omega) \cap L_{\Uparrow I})_{\Downarrow I} = (Pref(C^\omega))_{\Downarrow I} = (i_1 + i_2)^\star i_1^\star + i_2^\star = (i_1 + i_2)^\star = Pref(S^\omega)$. Moreover, L is the largest solution, because L is the universal language on alphabet U.

However, $\overrightarrow{L} = (u_1 + u_2)^\omega$ is not a solution of $C^\omega \diamond_\omega X \subseteq S^\omega$. Indeed, while $u_2^\star \in L$ is a solution of $Pref(C^\omega) \diamond X \subseteq Pref(S)^\omega$, the word $u_2^\omega \in \overrightarrow{L}$ is not a solution of the equation $C^\omega \diamond_\omega X \subseteq_\omega S^\omega$.

The reason is that the word $(i_2u_2)^\omega \in (u_2^\omega)_{\Uparrow \omega I}$ (the ω-expansion of u_2^ω to alphabet I), the word $(i_2u_2)^\omega \in C^\omega \cap_\omega ((u_1 + u_2)^\omega)_{\Uparrow \omega I}$ (the ω-intersection of C^ω and $((u_1 + u_2)^\omega)_{\Uparrow \omega I}$), and the word $i_2^\omega = ((i_2u_2)^\omega)_{\Downarrow \omega I} \notin S^\omega$ ($((i_2u_2)^\omega)_{\Downarrow \omega I}$ is the ω-restriction of $(i_2u_2)^\omega$ to alphabet I).

In this section we describe how to extend parallel equations to ω-languages described by Büchi automata, following the exposition in [24] and in the Master's Thesis by V. Bushkov to which we are indebted [23]. A similar blueprint could be followed to extend synchronous equations to ω-languages described by Büchi automata, but we are not going to work it out here.

4.2.1 Expansion and Restriction of ω-Languages

Definition 4.2.1. Given the disjoint alphabets A and B, an ω-language L defined over A, and the mapping $\psi : A^\omega \rightarrow 2^{(A \cup B)^\omega}$ such that $\psi(\alpha) =$

$\{u_1\alpha_1u_2\alpha_2\ldots u_i\alpha_i\cdots : u_i \in B^\star, i \in N\}$, the **$\omega$-expansion** of L to alphabet B (also called B-expansion of L) is the ω-language

$$L_{\Uparrow\omega B} = \cup_{\alpha\in L}\psi(\alpha),$$

i.e., in order to expand L to B we insert all finite words over B before and after any symbol α_i of any word $\alpha \in L$.

The generalization of the restriction operator from finitary languages to ω-languages is less straightforward because it must return an ω-language, even when the words of the initial ω-language end with ω-sequences of symbols only from alphabet B.

Definition 4.2.2. Given the disjoint alphabets A and B, an ω-language L defined over $A \cup B$,, and the mapping $h : A \cup B \to A$ such that $h(a) = a$ if $a \in A$ and $h(a) = \epsilon$ if $a \notin A$ (and by induction $h(\alpha) = h(\alpha_1)h(\alpha_2)\ldots h(\alpha_i)\ldots$, where $\alpha = \alpha_1\alpha_2\ldots\alpha_i\ldots \in L$), the **$\omega$-restriction** of L to alphabet A (also called A-restriction of L) is the ω-language

$$L_{\Downarrow\omega A} = \begin{cases} \emptyset & \text{if } L \subseteq (A\cup B)^\star B^\omega \\ \cup_{\alpha\in L\setminus(A\cup B)^\star B^\omega}\, h(\alpha) & \text{if } L \not\subseteq (A\cup B)^\star B^\omega \end{cases}$$

The definition of ω-restriction implies that if $L = \emptyset$ then $L_{\Downarrow\omega A} = \emptyset$.

Proposition 4.14. *Given the disjoint alphabets A and B, and the ω-language $L \subseteq A^\omega$, then $(L_{\Uparrow\omega B})_{\Downarrow\omega A} = L$.*

Proposition 4.15. *Given the disjoint alphabets A and B, and the ω-languages $L_1 \subseteq (A\cup B)^\omega$, $L_2 \subseteq (A\cup B)^\omega$ such that $(L_{2\Downarrow\omega A})_{\Uparrow\omega B} = L_2$, then $(L_1\cap L_2)_{\Downarrow\omega A} = L_{1\Downarrow\omega A} \cap L_{2\Downarrow\omega A}$.*

4.2.2 Parallel Composition of ω-Languages

Definition 4.2.3. Given the pairwise disjoint alphabets I, U, V and O, the ω-languages $L_1 \subseteq (I\cup U\cup V)^\omega$, $L_2 \subseteq (O\cup U\cup V)^\omega$, and a subset $E \subseteq I\cup V\cup O$, the **$\omega$-parallel composition**, denoted by $\diamond_{\omega E}$ of L_1 and L_2 is the ω-language

$$L_1 \diamond_{\omega E} L_2 = (L_{1\Uparrow O} \cap L_{2\Uparrow I})_{\Downarrow E}.$$

We use $\diamond_E$ or $\diamond_\omega$ instead of $\diamond_{\omega E}$ when the meaning is clear from the context. A parallel composition models that the components interact by means of common actions from $U \cup V$, i.e., an action from $U \cup V$ is executed if and only if both components are ready to execute it at their current states, while external actions from $I \cup O$ are executed independently.

Proposition 4.16. *If $L_2' \subseteq L_2$, then $L_1 \diamond_{\omega E} L_2 \subseteq L_1 \diamond_{\omega E} L_2$.*

4.2.3 Parallel Equations Over ω-Languages

Definition 4.2.4. Given the pairwise disjoint alphabets I, U, V and O, the ω-languages $C \subseteq (I \cup U \cup V)^\omega$, $S \subseteq (I \cup V \cup O)^\omega$, and the sets $E = I \cup V \cup O$ and $H = U \cup V \cup O$, $C \diamond_{\omega E} X \subseteq S$ (or $C \diamond_{\omega E} X = S$) is a **$\omega$-parallel equation**, with respect to the unknown ω-language X over alphabet H.

An ω-language $B \subseteq H^\omega$ is a **solution** of the equation $C \diamond_{\omega E} X \subseteq S$ (or $C \diamond_{\omega E} X = S$) if $C \diamond_{\omega E} B \subseteq S$ (or $C \diamond_{\omega E} B = S$).

$C \diamond_{\omega E} B \subseteq S$ (or $C \diamond_{\omega E} B = S$) is **solvable** if a solution exists.

A solution is called the **largest** solution, if it contains every solution.

In [24] the following statement is proved.

Theorem 4.17. *A solvable equation $C \diamond_{\omega E} X \subseteq S$ over ω-languages has the largest solution $\overline{(C \diamond_{\omega H} \overline{S})}$, where $\overline{S}$ denotes the complement of S.*

If $C \diamond_{\omega E} \overline{(C \diamond_{\omega H} \overline{S})} = S$, then $\overline{(C \diamond_{\omega H} \overline{S})}$ is the largest solution of the equation $C \diamond_{\omega E} X = S$.

Any ω-solution of the equation $C \diamond_{\omega E} B = S$ is contained in $\overline{(C \diamond_{\omega H} \overline{S})}$, but not every ω-language contained in $\overline{(C \diamond_{\omega H} \overline{S})}$ is a solution of the equation.

As done for finitary languages, solving an equation over ω-regular languages reduces to transforming the operator over ω-languages to operations over the corresponding ω-automata that accept them.

4.2.4 Operations Over ω-Automata

4.2.4.1 ω-Expansion of ω-Automata

Proposition 4.18. *Given a Büchi automaton $\mathcal{A} = (Q, A, \Delta, q_0, F)$ accepting a given ω-regular language L over the alphabet A and an alphabet B disjoint from A, consider the Büchi automaton $\mathcal{A}_{\Uparrow \omega B}$, derived from $\mathcal{A}$ as follows:*

for every state $s \in Q \setminus F$ and every symbol $b \in B$, add a self-loop (s, b, s); for every state $s \in F$, add a non-accepting state s', a transition (s, b, s'), a self-loop (s', b, s') under every symbol $b \in B$ and a transition (s', a, s'') for every transition (s, a, s'') where.

Then $\mathcal{A}_{\Uparrow \omega B}$ accepts the B-expansion $L_{\Uparrow \omega B}$ of the language L accepted by $\mathcal{A}$.

Example 4.19. Given the automaton $\mathcal{A}$ over $A = \{a, c\}$ shown in Fig. 4.3a, b shows $\mathcal{A}_{\Uparrow \omega B}$, where $B = \{b\}$. Notice that the operators for standard expansion of finitary languages and ω-expansion act differently on the automata. For instance, for ω-expansion we do not add self-loops at accepting states, otherwise $L(\mathcal{A}_{\Uparrow \omega B})$ would not be equal to $(L(\mathcal{A}))_{\Uparrow \omega B}$.

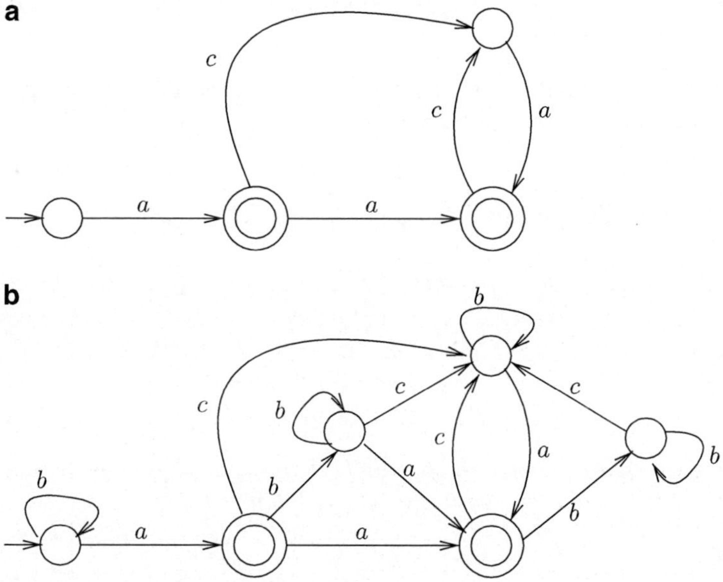

Fig. 4.3 (**a**) Automaton $\mathcal{A}$; (**b**) Automaton $(\mathcal{A})_{\Uparrow\omega B}$

4.2.4.2 ω-Restriction of ω-Automata

Proposition 4.20. *Given a Büchi automaton* $\mathcal{A} = (Q, A \cup B, \Delta, q_0, F)$ *accepting a given ω-regular language L over the alphabet* $A \cup B$*, consider the Büchi automaton* $\mathcal{A}_{\Downarrow\omega A}$*, derived from* $\mathcal{A}$ *by replacing every transition* (s, b, s') *where* $s \in Q$ *and* $b \in B \setminus A$ *by the transition* (s, ϵ, s')*, and then taking its ε-closure.*

Then $\mathcal{A}_{\Downarrow\omega A}$ *accepts the A-restriction* $L_{\Downarrow\omega A}$ *of the language L accepted by* $\mathcal{A}$*.*

4.2.4.3 ω-Product of ω-Automata

Proposition 4.21. *Given two Büchi automata* $\mathcal{A}_1 = (Q_1, A, \Delta_1, q_{01}, F_1)$ *and* $\mathcal{A}_2 = (Q_2, A, \Delta_2, q_{02}, F_2)$ *accepting respectively the given ω-regular languages* L_1 *and* L_2 *over the alphabet A, consider the Büchi automaton*

$$\mathcal{A}_1 \cap_\omega \mathcal{A}_2 = ((Q_\cap, A, \Delta_\cap, q_{0\cap}, F_\cap)),$$

defined as follows:

$$Q_\cap = Q_1 \times Q_2 \times \{1, 2\}; q_{0\cap} = (q_{01}, q_{02}, 1); F_\cap = Q_1 \times Q_2 \times \{2\};$$

$\Delta_\cap$ *is such that*

$((s_1, s_2, 1), a, (s_1', s_2', 1)) \in \Delta_\cap$ *iff* $(s_1, a, s_1') \in \Delta_1, (s_2, a, s_2') \in \Delta_2, s_1 \notin F_1,$
$((s_1, s_2, 1), a, (s_1', s_2', 2)) \in \Delta_\cap$ *iff* $(s_1, a, s_1') \in \Delta_1, (s_2, a, s_2') \in \Delta_2, s_1 \in F_1,$
$((s_1, s_2, 2), a, (s_1', s_2', 2)) \in \Delta_\cap$ *iff* $(s_1, a, s_1') \in \Delta_1, (s_2, a, s_2') \in \Delta_2, s_2 \notin F_2,$
$((s_1, s_2, 2), a, (s_1', s_2', 1)) \in \Delta_\cap$ *iff* $(s_1, a, s_1') \in \Delta_1, (s_2, a, s_2') \in \Delta_2, s_2 \in F_2.$

Then $\mathcal{A}_1 \cap_\omega \mathcal{A}_2$ accepts the intersection of the ω-languages accepted respectively by $\mathcal{A}_1$ and $\mathcal{A}_2$.

Notice the major differences between the operators for standard intersection of finitary languages and ω-intersection.

4.2.4.4 ω-Parallel Composition of ω-Automata

Proposition 4.22. *Given two Büchi automata $\mathcal{A}_1 = (Q_1, I \cup U \cup V, \Delta_1, q_{01}, F_1)$ and $\mathcal{A}_2 = (Q_2, U \cup V \cup O, \Delta_2, q_{02}, F_2)$ accepting respectively the given ω-regular languages L_1 and L_2, and the alphabet $E \subseteq I \cup V \cup O$, consider the Büchi automaton*

$$\mathcal{A}_1 \diamond_{\omega E} \mathcal{A}_2 = ((\mathcal{A}_1)_{\Uparrow \omega O} \cap_\omega (\mathcal{A}_2)_{\Uparrow \omega I})_{\Downarrow \omega E}.$$

Then $\mathcal{A}_1 \diamond_{\omega E} \mathcal{A}_2$ accepts the parallel composition of the ω-languages accepted respectively by $\mathcal{A}_1$ and $\mathcal{A}_2$, with respect to alphabet E.

4.2.4.5 ω-Parallel Complementation of ω-Automata

We omit the procedure for deriving the Büchi automaton that accepts the complement of the language accepted by a given Büchi automaton, because it is quite complicated. For a discussion on the topic we refer to [42, 53, 122].

The problem of complementation may be simplified in special cases, as done in Chap. 18, where a subset construction is used to obtain a deterministic Büchi over-approximation of an ND Büchi automaton. Therefore, the final complementation, done by simply complementing the acceptance conditions to obtain a co-Büchi automaton, yields a subset of the most general solution automaton.

4.2.4.6 Summing it up

The following theorem has been proved in [24].

Theorem 4.23. *Given the pairwise disjoint alphabets I, U, V and O, the ω-regular languages $C \subseteq (I \cup U \cup V)^\omega$, $S \subseteq (I \cup V \cup O)^\omega$, and the sets $E = I \cup V \cup O$ and $H = U \cup V \cup O$, suppose that $\mathcal{A}_C$ and $\mathcal{A}_S$ are Büchi automata accepting respectively C and S, then the largest solution of the ω-regular parallel equation*

$$C \diamond_{\omega E} X \subseteq S$$

is the ω-regular language accepted by the Büchi automaton

$$\overline{\mathcal{A}_C \diamond_{\omega H} \overline{\mathcal{A}_S}}.$$

If $C \diamond_{\omega E} L(\overline{\mathcal{A}_C \diamond_{\omega H} \overline{\mathcal{A}_S}}) = S$, *then* $L(\overline{\mathcal{A}_C \diamond_{\omega H} \overline{\mathcal{A}_S}})$ *is the largest solution of the equation* $C \diamond_{\omega E} X = S$.

We will describe in Chap. 18 an heuristic algorithm to solve synchronous equations over ω-languages represented by Büchi automata.

Problems

4.1. Show that the set of finite subsets of natural numbers, $\omega = \{0, 1, 2, \ldots\}$, represented as a set of strings in $\{0, 1\}^\omega$, is accepted by a non-deterministic Büchi automaton, but by no deterministic Büchi automaton (from [73]).

Notice that a set $A \subseteq \omega$ is represented by its characteristic string, i.e., the infinite string over $\{0, 1\}$ with a 1 in position i if and only if $i \in A$. For example the characteristic string of the set of multiples of 3 is:

100100100100100100100100...

the characteristic string of primes is:

0011010100010100010 10...

the characteristic set of singleton 4 is:

000010000000000000000...

4.2. Show that every set accepted by a Büchi automaton is a finite union of sets of the form AB^ω, where A and B are regular sets.

4.3. Consider the automaton in Fig. 4.4.

(a) What is its language when interpreted as a finite automaton?
(b) What is its language when interpreted as a Büchi automaton, with A as the final state?
(c) Is there a deterministic Büchi automaton to recognize the complement of the language defined in (b)?
(d) What is its language when interpreted as a co-Büchi automaton, with B as the persistent state?

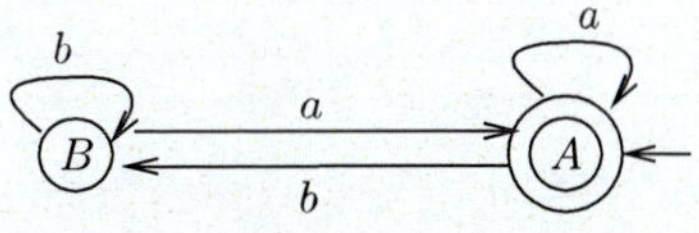

Fig. 4.4 (**a**) Büchi automaton for Problem 4.3

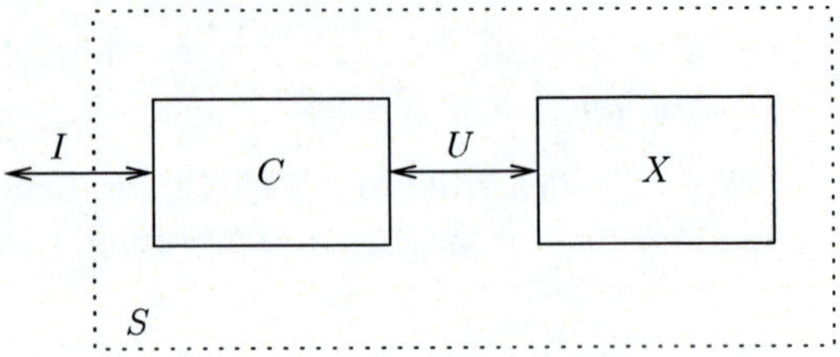

Fig. 4.5 Topology for Problem 4.4 and Problem 4.5

4.4. Given the topology shown in Fig. 4.5, consider the ω-regular languages $C^\omega = (i_1u_1 + i_1u_2 + i_2u_1 + i_2u_2)^\star(i_1u_1)^\omega + (i_2u_2)^\omega$ and $S^\omega = (i_1 + i_2)^\star i_1^\omega$.

Check whether the largest solution of the equation $C^\omega \bullet_\omega X \subseteq_\omega S^\omega$ is the limit of the largest solution of the equation $Pref(C^\omega) \bullet X \subseteq Pref(S^\omega)$. Notice that the former is a parallel equation over ω-regular languages and the latter is a parallel equation over regular languages (obtained as finite prefixes of ω-languages).

4.5. Given the topology shown in Fig. 4.5, consider the ω-regular languages $C^\omega = ((i_1 + i_2)(u_1 + u_2))^\omega$ and $S^\omega = (i_1 + i_2)^\omega$.

Check whether the largest solution of the equation $C^\omega \diamond_\omega X \subseteq_\omega S^\omega$ is the limit of the largest solution of the equation $Pref(C^\omega) \diamond X \subseteq Pref(S^\omega)$. Notice that the former is a parallel equation over ω-regular languages and the latter is a parallel equation over regular languages (obtained as finite prefixes of ω-languages).

Chapter 5
A Survey of Relevant Literature

5.1 Equations Under Synchronous Composition

Sequential synthesis offers a collection of problems that can be modeled by FSM equations under synchronous composition. Some have been addressed in the past with various techniques in different logic synthesis applications.

5.1.1 Hierarchical Optimization and Don't Care Sequences

In place of designing a huge monolithic FSM and then optimizing it by state reduction and encoding, it is convenient to work with a network of smaller FSMs. However, if each of them is optimized in isolation, part of the implementation flexibility is lost, because no use is made of the global network information. Hierarchical optimization calls for optimizing the FSMs of a network capturing the global network information by means of don't care conditions. The goal of hierarchical optimization is to optimize the FSMs of a network capturing the global network information by means of don't care conditions. This paradigm follows the approach taken in multi-level combinational synthesis since the beginning [27, 29], where a lot of effort has been invested in capturing the don't cares conditions and devising efficient algorithms to compute them or their subsets [44, 54, 57, 95].

The first source of don't care sets is due to external limitations on controllability and observability. In particular *input controllability don't cares* and *output observability don't cares* have been defined for multi-level combinational networks (see [57, 95]).

Definition 5.1.1. The **input controllability don't care set**, CDC_{in}, includes all input vectors that are never produced by the environment at the network's inputs. The **output observability don't care sets**, ODC_{out}, denote all input vectors that represent situations when an output is not observed by the environment.

T. Villa et al., *The Unknown Component Problem: Theory and Applications*,
DOI 10.1007/978-0-387-68759-9_5, © Springer Science+Business Media, LLC 2012

When the theory is extended to sequential circuits, don't care sets become sequences of inputs instead of single inputs, since sequential circuits transform input sequences into output sequences.

Definition 5.1.2. The **input controllability sequential don't care set,** CDC_{in}^{seq}, includes all input sequences that are never produced by the environment at the network's inputs. The **output observability sequential don't care sets,** ODC_{out}^{seq}, denote all input sequences that represent situations when an output is not observed by the environment at the current time or in the future, and all subsets of input sequences whose outputs are not distinguished by the environment.

When the previous definitions are applied to the components of a network, the limited controllability or observability becomes a consequence of what cannot be produced by a source component or cannot be observed/distinguished by a sink component. These restrictions can be exploited to optimize the components; for instance, since the output don't care sequences of a component are the subsets of input sequences that cannot be distinguished at the outputs of the component, they can be used to modify the component that produces them since the modifications carry no effect on the component that receives them as inputs.

In the rest of the chapter, we will represent sometimes a composition like $M_A \bullet M_B$ with the notation $M_A \to M_B$ or $M_A \leftrightarrow M_B$, to highlight that it is a series composition ($M_A \to M_B$) vs. a closed-loop ($M_A \leftrightarrow M_B$).

5.1.2 *Computation of Input Don't Care Sequences*

Consider a cascade interconnection of two FSMs M_A and M_B, where the driving FSM M_A feeds the input vectors to the driven FSM M_B. Then input controllability don't cares are the sequences of outputs not produced by M_A; they restrict the controllability of the driven FSM M_B and are used to modify M_B obtaining an FSM $\hat{M}_B$ such that the cascade interconnection does not change, i.e., $M_A \to \hat{M}_B = M_A \to M_B$.

Kim and Newborn [69] were the first to give a procedure to compute all input controllability don't care sequences for a series topology, as follows:

1. Construct a NDFA $NA^o_{M_A}$ accepting the output language produced by the FSM M_A. $NA^o_{M_A}$ is derived from M_A by replacing the input/output labels by the output labels and making each state acceptable; moreover a dead state F is added to $NA^o_{M_A}$, and if from state s there is no transition with label o then a transition from s to F under o is added to $NA^o_{M_A}$.
2. Obtain $A^o_{M_A}$ as the determinization of $NA^o_{M_A}$ (and minimize its states, if desired).

3. Compute $\hat{M}_B$ as the product of $A^o_{M_A}$ and M_B,[1] except that transitions going into states of the product machine containing F are considered as unspecified transitions (in other words, they are sent to a dummy state DC with a self-loop under every input asserting every output). So $\hat{M}_B$ is an ISFSM that can be reduced using conventional algorithms for state minimization. After minimization, $\hat{M}_B$ is guaranteed to have no more states than M_B.

In [121] a heuristic was provided to reduce the number of states of $A^o_{M_A}$ by discarding some don't care information; still $A^o_{M_A}$ has to be built first. More heuristics to approximate input don't care sequences were explored in [139].

Later on, Wang and Brayton [139] showed that input don't care sequences for a component in a network of FSMs with an arbitrary topology can be exploited in the same way as in a series topology and that computing input don't care sequences for an arbitrary topology can be reduced to computing them for a series topology. Given the FSMs M_A and M_B composed in a two-way topology, $M_A \leftrightarrow M_B$, it was proved that if $\overline{\mathcal{L}_x}$ are the input don't cares sequences for M_B, and if M_B' is obtained from M_B using $\overline{\mathcal{L}_x}$ as unspecified transitions and M_B'' is obtained from M_B' after state minimization, it holds that $M_A \leftrightarrow M_B = M_A \leftrightarrow M_B''$, i.e., the behavior of the network is preserved when M_B'' replaces M_B.

An FSM network with an arbitrary topology and including FSM M_B can be reduced to the case of a two-way topology, by lumping together as FSM M_A all the FSM components except for M_B. In this way, the original network is reduced to either a series topology or a two-way topology of FSMs M_A and M_B. As a result, input don't care sequences for a component in a general FSM network can be exploited using state minimization procedures for incompletely specified FSMs. To compute the input don't care sequences with the procedure of Kim and Newborn, a further step is needed to reduce from a two-way topology to a series topology, as follows: consider a network $M_A \leftrightarrow M_B$ and call it $\mathcal{N}$, then network $\mathcal{N}$ produces the same I/O sequences as network $\mathcal{N}'$ drawn in Fig. 5.1, because $M_A \leftrightarrow M_B$ produces the same U and V sequences in $\mathcal{N}$ as it does in $\mathcal{N}'$, therefore M_A and M_B behave under U and V in network $\mathcal{N}'$ as they do in $\mathcal{N}$, and so $\mathcal{N}'$ produces the same I/O sequences as network $\mathcal{N}$. By this transformation from a two-way topology to a series topology, the input don't care sequences respectively of M_A and M_B are preserved. The driving machine is in both cases the composition of FSMs M_A and M_B, $M_A \leftrightarrow M_B$, and it is called the abstract driving machine of the FSM network.

In general, given an FSM network and a component FSM M_B whose input don't care sequences should be computed, the abstract driving machine is the network itself, unless M_B is in a one-way communication with the other components and so the abstract driving machine is the network except M_B. The previous steps establish that it is correct to apply the procedure by Kim and Newborn directly to an arbitrary FSM network topology. In summary:

[1] One has to take care of defining over the same alphabets both $A^o_{M_A}$ and M_B, for instance turning $A^o_{M_A}$ into an FSM with don't care outputs for each input combination of the automaton $A^o_{M_A}$.

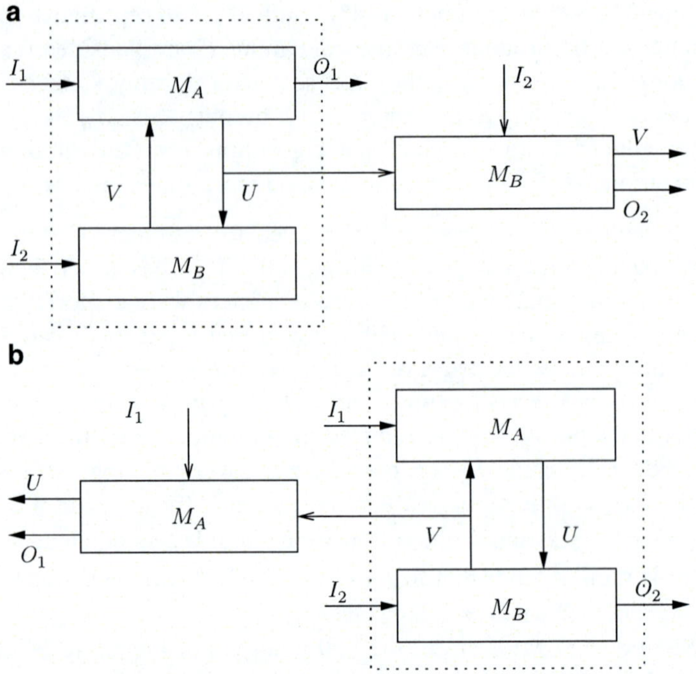

Fig. 5.1 Networks $\mathcal{N}'$ obtained by series composition of driving machine $\mathcal{N} = M_A \leftrightarrow M_B$ with M_B or M_A. (**a**) The series topology $(M_A \bullet M_B)_{\downarrow I_1 \times I_2 \times U \times O_1} \rightarrow M_B$ is used to compute the input don't care sequences of M_B; (**b**) The series topology $(M_A \bullet M_B)_{\downarrow I_1 \times I_2 \times V \times O_2} \rightarrow M_A$ is used to compute the input don't care sequences of M_A

1. The series topology $(M_A \bullet M_B)_{\downarrow I_1 \times I_2 \times V \times O_2} \rightarrow M_B$ preserves the input don't care sequences of M_B.
2. The series topology $(M_A \bullet M_B)_{\downarrow I_1 \times I_2 \times U \times O_1} \rightarrow M_A$ preserves the input don't care sequences of M_A.

Example 5.1. Figure 5.2 shows a series topology $M_A \rightarrow M_B$ and the computation of the input don't care sequences with the procedure by Kim and Newborn, providing the full flexibility at M_B. Notice that $\hat{M}_B$ can be reduced to two states (by merging states 22, 33 and $3T$), whereas M_B has three states.

The theory of input don't care sequences was developed independently in the Soviet Union by Yevtushenko [152].

Another approach based on injecting faults into the component FSM M_B and checking whether the injected fault is redundant was proposed by Ferrandi et al. in [40]. This method does not compute the complete flexibility, but the experiments reported show that the computed partial flexibility is sufficient to achieve a good optimization of M_B.

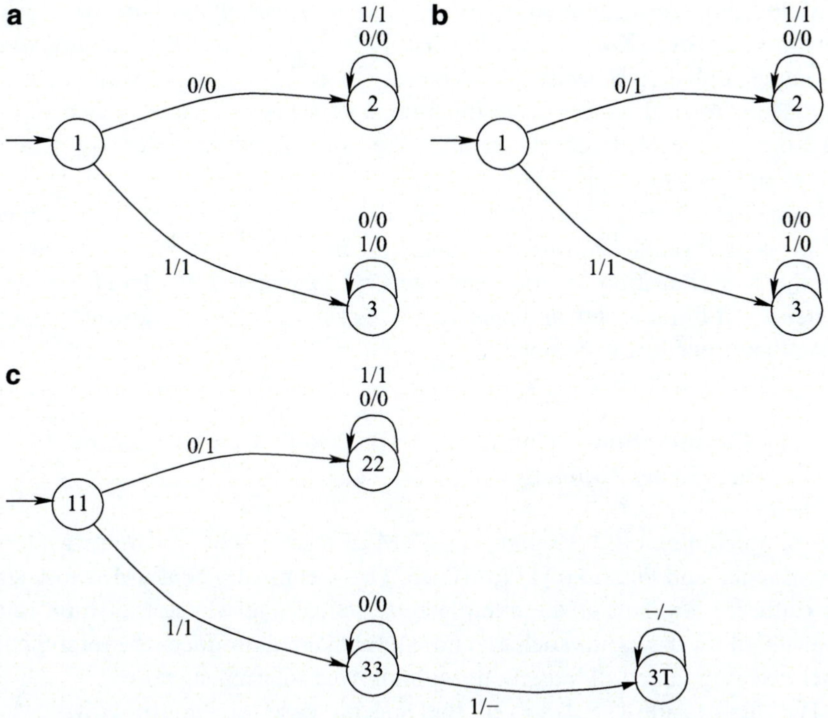

Fig. 5.2 Components (**a**) M_A and (**b**) M_B of series topology $M_A \rightarrow M_B$ of Example 5.1; (**c**) $\hat{M}_B$ expresses the full flexibility at M_B due to input don't care sequences; $\hat{M}_B$ can be reduced to two states by state minimization

5.1.3 Computation of Output Don't Care Sequences

Output observability don't cares are the sets of sequences of inputs of M_B that cannot be distinguished by the outputs of M_B, i.e., the sequences of a set cannot be distinguished from each other by looking to the outputs of M_B; they restrict the observability of the driving FSM M_A and are used to modify M_A obtaining an FSM $\hat{M}_A$ such that the cascade interconnection does not change, i.e., $\hat{M}_A \rightarrow M_B = M_A \rightarrow M_B$.

5.1.3.1 Computation of Subsets of Output Don't Care Sequences

Output don't cares sequences of length one (called simply sequential output don't cares), were studied by Devadas [34], and fixed-length output don't care sequences by Rho [121].

A different subset of output don't care sequences of M_A is computed by a procedure designed by Wang [140], whose idea is to enumerate implicitly all possible languages generated by M_A *by changing the output function, but keeping the state transition function unchanged.* Say that M_A has k transition edges and an output alphabet U with $|U|$ symbols, then M_A may produce $|U|^k$ possible output functions, if its transition function is kept unchanged. For each such output function λ_1' one must check whether the corresponding FSM M_A' is such that $M_A' \to M_B = M_A \to M_B$.

Comparing Wang's and Rho's procedures, the former implicitly enumerates infinite-length output don't care sequences, but it is restricted by the assumption that the state transition function must be kept unchanged; the latter is restricted to sequences of finite-length and also uses a restricted notion of equivalent sequences. So neither procedure is complete.

5.1.3.2 Computation of Complete Output Don't Care Sequences in a Series Topology

The complete flexibility for the head FSM of a series composition was derived by Yevtushenko and Petrenko [111, 112, 152] by means of a NDFSM whose states are the Cartesian product of the components' states, and whose transition relation is unspecified for the inputs such that no internal signal produces the reference output, otherwise it includes all transitions with allowed internal signals.

The first result [152] solved the special case of Moore FSMs where the tail component produces different outputs for different states. Consider a series composition $M_A \to M_B$ of two Moore FSMs $M_A = (S_A, I, U, \delta_A, \lambda_A, r_A)$ and $M_B = (S_B, U, O, \delta_B, \lambda_B, r_B)$, such that $\forall s_1, s_2 \in S_B \; s_1 \neq s_2$ implies $\lambda_B(s_1) \neq \lambda_B(s_2)$. The FSM representing all behaviors that can be realized at the head component is given by the NDFSM $M_D = (S_A \times S_B, I, U, \delta_D, \lambda_D, (r_A, r_B))$, where $\delta_D((s_A, s_B), i) = (\delta_A(s_A, i), \delta_B(s_B, \lambda_A(s_A)))$ and $\lambda_D(s_A, s_B) = \{u \mid \delta_B(s_B, y) = \delta_B(s_B, \lambda_A(s_A))\}$, i.e., the output of M_D at state (s_A, s_B) is the set of $u \in U$ that drive M_2 from s_B into the same state to which $\lambda_A(s_A)$ does.

Theorem 5.2. *[152] $M_C \to M_B = M_A \to M_B$ iff M_C is a reduction of M_D.*

The method was then extended to arbitrary tail machines through two more contributions [111, 112]. The first one described in [112] proposes an algorithm for output don't care sequences dual to the one by Kim and Newborn for input don't care sequences.

Consider a series composition $M_A \to M_B$ of two FSMs $M_A = (S_A, I, U, \delta_A, \lambda_A, r_A)$ and $M_B = (S_B, U, O, \delta_B, \lambda_B, r_B)$. The FSM representing all classes of input sequences of M_B equivalent with respect to M_B (M_B produces the same output sequence under these input sequences, which are the don't care output sequences of M_A) is given by the NDFSM $M_D = (S_B \times S_B, U, U, T, (r_B, r_B))$, where the transition $((\hat{s}_B, \tilde{s}_B), u_1, u_2, (\hat{s}'_B, \tilde{s}'_B)) \in T$ iff $\lambda_B(\hat{s}_B, u_1) = \lambda_B(\tilde{s}_B, u_2)$, $\hat{s}'_B = \delta_B(\hat{s}_B, u_1)$ and $\tilde{s}'_B = \delta_B(\tilde{s}_B, u_2)$. In other words, the output sequences produced by FSM M_D under a given input sequence α are those under which M_B produces the same output sequence that it generates under α.

Theorem 5.3. *[112] $M_C \rightarrow M_B = M_A \rightarrow M_B$ iff M_C is a reduction of the product of M_A and M_D.*

The contribution described in [111] builds directly the NDFSM capturing all the flexibility.[2]

Example 5.2. Figure 5.3a,b show the head and tail components of a series topology $M_A \rightarrow M_B$ to illustrate the computation of the output don't care sequences with the procedure by Yevtushenko and Petrenko in [111], providing the full flexibility at M_A. Figure 5.3c portrays the NDFSM M_D representing all classes of input sequences equivalent with respect to M_B (M_B produces the same output sequence under these input sequences, which are the don't care output sequences of M_A). Then the product $M_A \bullet M_D$ contains all the behaviors that can replace the given head component M_A. One such behavior, corresponding to a submachine M'_A of $M_A \bullet M_D$ is shown in Fig. 5.3d, and it can be reduced to two states by state minimization as in Fig. 5.3e (states a_1, c_2, c_4 are merged into one state, and states b_1, b_4 into another), whereas the original head machine M_A had three states.

5.1.4 Computation of the Permissible Behaviors with the E-Machine

The attempts to characterize with input and output don't care sequences the complete flexibility of a component in a composition with loops failed, because modifying a component affects in a circular way the input and output don't care sequences computed so far.

The first description of the complete flexibility in a two-way synchronous composition for the rectification topology is due to Y. Watanabe and R. Brayton. Given the network topology shown in Fig. 1.1d, they introduced in [142] a fixed-point computation to compute a PNDFSM that contains *all* behaviors M_B (DFSMs) whose composition with the given machine M_A is contained in the specification M_C. The PNDFSM so obtained has been called the *E-machine*, where the prefix *E* stands for environment. An alternative computation, credited to A. Saldanha, builds an equivalent NDFSM (see [66] for a detailed exposition). The authors have also investigated the issue of logical implementability of the DFSMs contained in the E-machine, i.e., the problem of finding those contained DFSMs M_B such that there

[2]Consider a series composition $M_A \rightarrow M_B$ of two FSMs $M_A = (S_A, I, U, \delta_A, \lambda_A, r_A)$ and $M_B = (S_B, U, O, \delta_B, \lambda_B, r_B)$. The FSM representing all behaviors that can be realized at the head component is given by the NDFSM $M_D = (S_A \times S_B \times S_B, I, U, T, (r_A, r_B, r_B))$, where $((s_A, \hat{s}_B, \tilde{s}_B), i, u, (s'_A, \hat{s}'_B, \tilde{s}'_B)) \in T$ iff the output of $M_A \rightarrow M_B$ at state $(s_A, \hat{s}_B)$ under input i is equal to the output of M_B at state $\tilde{s}_B$ under input u, and $(s'_A, \hat{s}'_B, \tilde{s}'_B)$ are the successor states respectively in $M_A \rightarrow M_B$ and M_B.

Theorem 5.4. *[111] $M_C \rightarrow M_B = M_A \rightarrow M_B$ iff M_C is a reduction of M_D.*

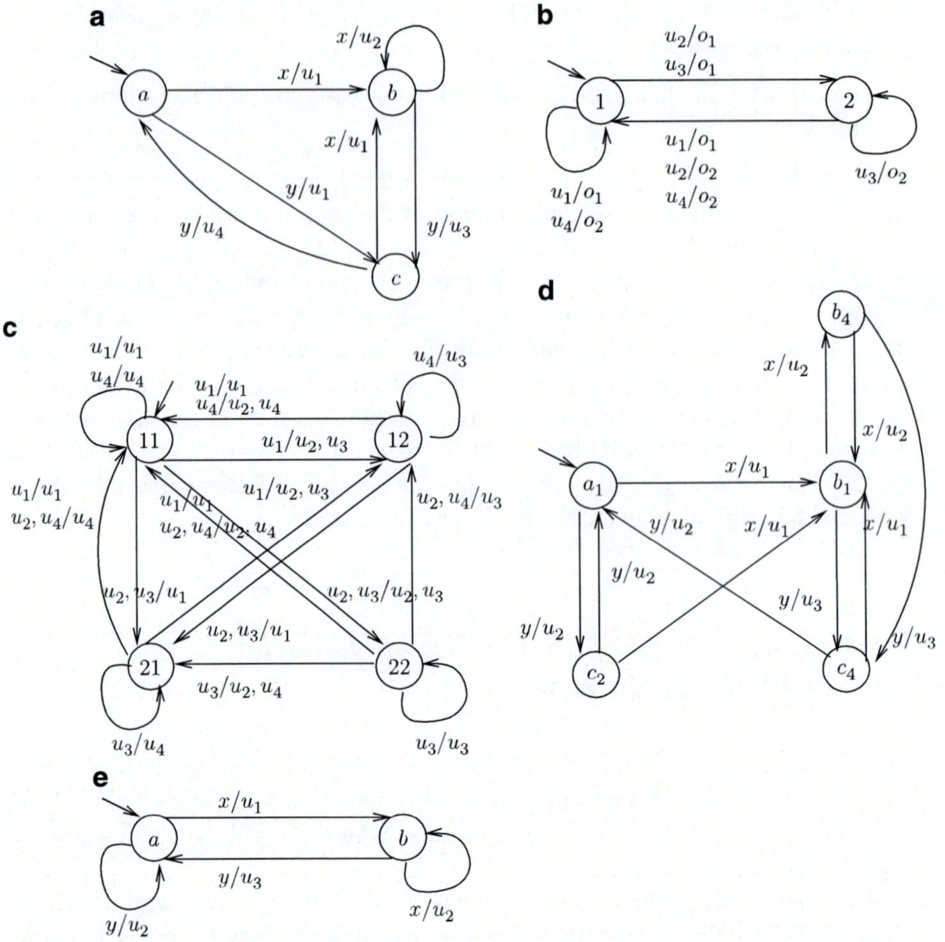

Fig. 5.3 Components (**a**) Head FSM M_A and (**b**) Tail FSM M_B of series topology $M_A \to M_B$ of Example 5.2; (**c**) FSM M_D that captures the flexibility at M_A due to output don't care sequences; (**d**) Submachine M'_A extracted from $M_A \bullet M_D$; (**e**) FSM M''_A obtained by state minimization of M'_A

exists a pair of circuit implementations of M_B and M_A with no combinational cycles created by connecting them together by the internal signals u and v. These DFSMs are called *permissible*.

5.1.5 How to Exploit the Flexibility

A common way to exploit the computed flexibility is first to minimize the number of states of each legal FSM and then the size of the logic realization of the encoded FSM. State minimization of ISFSMs has been addressed in the classical literature starting from the seminal papers of the 1960s by Grasselli and Luccio [50, 65], and then extended in the 1990s to PNDFSMs [32, 67, 143]. A comprehensive survey on state minimization of FSMs can be found in [66]. Using state minimizers to exploit the flexibility has two drawbacks: (1) exact state minimization is a computationally hard problem; (2) reducing the number of states does not guarantee a smaller logic realization.

The problem of FSM encoding in the context of a network of FSMs is even more complex than for a monolithic FSM (because symbolic variables may be shared among components) and it has received up to now scant attention in the literature.

Optimization techniques for sequential circuits at the netlist level are in a more mature stage than state-based methods, since they achieve good results on large circuits; however, manipulating symbolic (state-based) information is indispensable for computing the flexibility of a component in an FSM network. To handle this issue, Wang proposed in [141] an algorithmic strategy which takes a circuit implementation as the starting point and computes the flexibility at the symbolic level, but exploitation is directly at the netlist logic level.

5.1.6 FSM Network Synthesis by WS1S

It is known since the seminal work of Büchi and Elgot in the sixties that regular languages play a role in decidability questions of formal logic [19]. For instance, it was shown that WS1S (Weak Second-Order Logic of 1 Successor) is decidable, because the set of satisfying interpretations of a subformula is represented by a finite automaton, and so the automaton for a formula can be constructed by induction, noticing that logical connectives correspond to operations on automata such as product, projection and subset construction. Validity, satisfiability or unsatisfiability of a formula can be established by answering questions on the corresponding automaton. So WS1S can be seen as another notation for regular languages, as also regular expressions are. WS1S features first-order variables that denote natural numbers (they can be compared and added to constants), and second-order variables that denote finite sets of numbers. We sketch next the basic steps that show the equivalence between WS1S logic and automata [70].

In WS1S any formula ϕ can be put in a form where atomic subformulas denoting either a subset, or a set difference, or a successor relation are operated upon by logical operators denoting either a negation, or a conjunction, or an existential quantification.

Given a formula ϕ, its semantics is defined inductively relative to a string w over the alphabet $\mathbb{B}^k$, where $\mathbb{B} = \{0, 1\}$ and k is the number of variables in ϕ. Assign to every variable of ϕ a unique index i in the range $1, 2, \dots, k$ and denote as P_i the variable whose index is i. The interpretation of P_i under the string w is the finite set $w(P_i) = \{j \mid P_i[j] = 1\}$, where $P_i[j] = 1$ denotes that the j-th bit in the P_i-th track is 1. The notation $w \models \phi$, read w satisfies ϕ, means that w makes ϕ true. The inductive construction follows the rules: $w \models \neg\phi$ iff $w \not\models \phi'$;

$w \models \phi' \wedge \phi''$ iff $w \models \phi'$ and $w \models \phi''$; $w \models \exists P_i\ \phi$ iff $\exists$ a finite $M \subseteq \mathbb{N}$ such that $w[P_i \mapsto M] \models \phi'$, where $w[P_i \mapsto M]$ denotes the shortest string w' that interprets all variables $P_j, i \neq i$, as w does, but interprets P_i as M; $w \models P_i \subseteq P_j$ iff $w(P_i) \subseteq w(P_j)$; $w \models P_i = P_j \setminus P_k$ iff $w(P_i) = w(P_j) \setminus w(P_k)$; $w \models P_i = P_j + 1$ iff $w(P_i) = \{k + 1 \mid k \in w(P_j)\}$.

Example 5.3. The formula $\phi = \forall p\ (P(p) \leftrightarrow \neg Q(p))$ defines the language $L(\phi)$ over $\mathbb{B}^2$ such that if the letter

$$\begin{vmatrix} x_1 \\ x_2 \end{vmatrix} \in \mathbb{B}^2$$

occurs in position j, then $j \in P$ iff $x_1 = 1$ and $j \in Q$ iff $x_2 = 1$. The string

$$w' = \begin{vmatrix} 0\,1\,1\,0 \\ 1\,0\,0\,1 \end{vmatrix} \in L(\phi),$$

whereas the string

$$w'' = \begin{vmatrix} 0\,1\,1 \\ 0\,0\,0 \end{vmatrix} \notin L(\phi).$$

For a formula ϕ define the language $\mathcal{L}(\phi)$ as the set of satisfying strings $\mathcal{L}(\phi) = \{w \models \phi\}$. Then the following result holds.

Theorem 5.7 (Thatcher-Wright [132]). *$L \subseteq (\{0,1\}^k)^\star$ is regular iff there exists a WS1S formula ϕ with $X_1, \ldots, X_k$ as free variables and $\mathcal{L}(\phi) = L$.*

To prove that any $\mathcal{L}(\phi)$ is a regular language, one shows by induction on the formula how to construct a deterministic automaton A such that $\mathcal{L}(A) = \mathcal{L}(\phi)$, where $\mathcal{L}(A)$ is the language recognized by A. The atomic formulas are translated into appropriate basic automata (see [71]). Composite formulas are handled by well-known automata operations: negation of a formula corresponds to automaton complementation (if $\phi = \neg\phi'$, then $\mathcal{L}(\neg\phi') = \overline{\mathcal{L}(\phi')} = \overline{\mathcal{L}(A')} = \mathcal{L}(\overline{A'})$, where A' is the automaton such that $\mathcal{L}(\phi') = \mathcal{L}(A')$, and so $A = \overline{A'}$ is the automaton for ϕ); conjunction corresponds to language intersection; existential quantification corresponds to projection, that may introduce non-determinism and require determinization by subset construction.

The converse statement that every regular language is representable in WS1S is proved by encoding in WS1S the finite automaton that recognizes the given regular language.

Example 5.8. [93] The nondeterministic automaton of Fig. 5.4 can be represented by the formula in WS1S

$$\forall t\ \ ([S_1(t) \wedge S_2(t) \wedge \neg I(t) \wedge S_1(t+1) \wedge S_2(t+1)]$$

$$\vee\, [S_1(t) \wedge S_2(t) \wedge \neg I(t) \wedge \neg S_1(t+1) \wedge S_2(t+1)]$$

Fig. 5.4 Nondeterministic automaton of Example 5.8

Input	Present state	Next state
b	k_0	k_0
b	k_0	k_2
a	k_1	k_0
a	k_2	k_0
a	k_2	k_1

$$\vee [S_1(t) \wedge \neg S_2(t) \wedge I(t) \wedge S_1(t+1) \wedge S_2(t+1)]$$
$$\vee [\neg S_1(t) \wedge S_2(t) \wedge I(t) \wedge S_1(t+1) \wedge \neg S_2(t+1)]$$
$$\vee [\neg S_1(t) \wedge S_2(t) \wedge I(t) \wedge S_1(t+1) \wedge S_2(t+1)])$$

where $I(t)$ means input a, $\neg I(t)$ means input b, $S_1(t) \wedge S_2(t)$ means state k_0, $S_1(t) \wedge \neg S_2(t)$ means state k_1 and $\neg S_1(t) \wedge S_2(t)$ means state k_2.

Even though the space and time requirements to translate formulas into automata have been shown to be bounded below by a stack of exponentials whose height is proportional to the length of the formula, an "efficient" implementation of the decision procedures of WS1S has been presented in the tool MONA [71], which computes a minimum deterministic automaton whose language is the set of strings for which the formula holds. Efficient means that a variety of interesting practical problems have been successfully handled by the program.

The WS1S formalism enables to write down easily equations that characterize the set of (functionally) permissible behaviors at a node for different topologies, as pointed out by Aziz et al. in [5]. We present the equations for some FSM networks shown in Fig. 1.1. In all equations M_C represents the specification; in a given equation, we label the unknown FSM by a superscript *. According to the WS1S syntax, the FSMs must be encoded to appear in the equations.

1-Way Cascade (a) - Fig. 1.1c $\phi^{M_A{}^*}(I_1, U) = (\forall O_2)[\phi^{M_B}(U, O_2) \rightarrow \phi^{M_C}(I_1, O_2)]$.
The machine M_A^* is exactly the one produced by the construction due to Kim and Newborn [69].

1-Way Cascade (b) - Fig. 1.1c $\phi^{M_B{}^*}(U, O_2) = (\forall I_1)[\phi^{M_A}(I_1, U) \rightarrow \phi^{M_C}(I_1, O_2)]$.

Supervisory Control - Fig. 1.1e $\phi^{M_B{}^*}(I_2, O_1, V) = \phi^{M_A}(V, O_1) \rightarrow \phi^{M_C}(I_2, O_1)$.
The restriction to Moore solutions was investigated in [4].

2-Way Cascade (a) - Fig. 1.1b $\phi^{M_A{}^*}(I_1, V, U) = (\forall O_2)[\phi^{M_B}(U, V, O_2) \rightarrow \phi^{M_C}(I_1, O_2)]$.

2-Way Cascade (b) - Fig. 1.1b $\phi^{M_B{}^*}(U, V, O_2) = (\forall I_1)[\phi^{M_A}(I_1, V, U) \rightarrow \phi^{M_C}(I_1, O_2)]$.

Rectification (a) - Fig. 1.1d $\phi^{M_B{}^*}(U, V) = (\forall I_1, O_1)[\phi^{M_A}(I_1, V, U, O_1) \rightarrow \phi^{M_C}(I_1, O_1)]$.

Rectification (b) - Fig. 1.1d $\phi^{M_A{}^*}(I_1, V, U, O_1) = \phi^{M_B}(U, V) \rightarrow \phi^{M_C}(I_1, O_1)$.

Going beyond previous *ad hoc* approaches, the fact of embedding logic synthesis problems into WS1S formulas allows one to state them in a common frame, enabling easily proof of correctness and completeness of the proposed solutions. Then the computations are performed on the related automata applying to them the operations that correspond to the standard logical connectives (however the result may be a regular language that is not an FSM language, an issue addressed by Theorem 3.8).

In contrast, the theory of synchronous and parallel equations is built upon the primitive notions of language and language composition, and models naturally a larger spectrum of language equations and their specialized solutions.

5.1.7 Testing

When an FSM is embedded into a complex system, usually there is no direct access to its inputs and outputs, and thus the FSM cannot be tested in isolation; the reason is limited controllability from outside of the internal inputs of the FSM, or limited observability from outside of the internal outputs of the FSM. This is the so-called problem of testing in context, which has been an active area of research. Petrenko et al. [114] proposed to solve testing in context by means of solving equations over FSMs to derive all FSMs externally equivalent to a given component FSM, since the latter FSMs are exactly the reductions of the largest solution describing the complete flexibility of the given component FSM.

5.1.8 Model Matching by Simulation Relations

The objective of model matching in control theory is to design a controller M_B so that the composition of a plant M_A with M_B matches a given model M_C (see the controller's topology in Fig. 1.1e). A procedure for deriving the largest solution for complete FSMs M_A (DFSM) and M_C (PNDFSM) of the equation $M_A \bullet M_X \preceq^{sim} M_C$ for the discrete model matching problem was proposed in [12, 68], where $\preceq^{sim}$ denotes simulation relation[3] (as opposed to language containment in

[3] $\psi \subseteq S_1 \times S_2$ is a **simulation relation** from an FSM $M_1 = \langle S_1, I, O, T_1, r_1 \rangle$ to an FSM $M_2 = \langle S_2, I, O, T_2, r_2 \rangle$ if:

1. $(r_1, r_2) \in \psi$, and
2. $(s_1, s_2) \in \psi \Rightarrow$

$$\{ \forall i \; \forall o \; \forall s_1' \; [(s_1 \xrightarrow{i/o}_{M_1} s_1') \Rightarrow \exists s_2' \; [(s_2 \xrightarrow{i/o}_{M_2} s_2') \wedge (s_1', s_2') \in \psi]] \}.$$

If such a ψ exists, we say that M_2 **simulates** M_1, or that M_1 has a simulation into M_2, and denote it by $M_1 \preceq^{sim} M_2$.

our approach). Simulation relations in general are stronger (more restrictive) than language containment, i.e., a simulation relation implies language containment, but not vice versa.

The use of simulation relation instead of language containment avoids determinization and leads to an algorithm of polynomial complexity bounded by $O(|S_A|.|S_C|.|T_A|.|T_C|)$, where $|S_A|$ ($|S_C|$) is the number of states of M_A (M_C) and $|T_A|$ (T_C) is the size of the transition relation of M_A (M_C). In [12] a solvability condition is given, based on the notion of simulation relation between the automata which generate the possible output sequences produced by an FSM. Model matching was investigated also within the frame of supervisory control in [8], relying on the relation between the controllability of a language and a given bisimulation; the same formulation was adopted in [91], where an implicit representation of automata based on algebraic methods is proposed.

Notice that for some topologies, like the rectification topology, solutions based on simulation relations do not provide the complete flexibility.

5.1.9 Structural Replacement of Synchronous and Asynchronous Hardware

We list a few lines of investigation covering topics that arise from structural replacement, when the implementability of a legal behavior is considered:

1. Model matching of asynchronous sequential machines. Issues of races, infinite cycles and adversarial inputs were discussed in [47, 99, 106, 133, 146].
2. Replacement and verification of hierarchical models of synchronous and asynchronous circuits. Trace theory has been used by Dill [35] to model hierarchical verification of asynchronous hardware. E. Wolf developed it further in [22, 145] to derive a closed-form expression that specifies all and only the permissible replacement circuitry in a given location in a circuit or partially-structured specified model. Her frame applies to both combinational circuits and clocked sequential circuits. Nondeterministic specifications are allowed too. The circuit semantics adopted allows the expression of circuits that contain unlatched feedback loops, and therefore, among the degrees of freedom available for the correct implementation of a part of a logic network, there is the possibility that it contains unlatched feedback.
3. Safe replaceability. It is the problem of replacing components when there is limited or no information on the initial states of the component under replacement, After the pioneering work of C. Pixley, important contributions came from Singhal and Brayton [58, 88, 127–129].

5.2 Equations Under Parallel Composition

Inequalities and equations under parallel composition have a wide range of applicability in various domains. For instance they found a fertile area of application in protocol converter synthesis [3, 6, 13, 103–105]. In this section we will survey some important related literature.

5.2.1 *Equations Over Process Languages and Process Algebras*

A process language is usually a prefix-closed regular language and is represented as the language (or set of traces) of a labeled transition system (LTS), which is a finite automaton (with ϵ moves) where each state is accepting. Some states can be marked also with a partially ordered set of marks.

In [94] the equation $A \diamond X = C$ was studied over LTS languages and the following results were claimed:

Theorem 5.9. *The general solution of $A \diamond X \subseteq C$ over prefix-closed regular languages is given by*

$$S = A \diamond C \setminus A \diamond \overline{C}.$$

Theorem 5.10. *The equation $A \diamond X = C$ is solvable over prefix-closed regular languages iff the language $S = A \diamond C \setminus A \diamond \overline{C}$ is a solution of the equation, i.e., $A \diamond (A \diamond C \setminus A \diamond \overline{C}) = C$.*

Theorem 5.9 does not compute the largest solution of $A \diamond X \subseteq C$ that is $S = \overline{A \diamond \overline{C}}$; the largest solution is obtained by adding to the language $A \diamond C \setminus A \diamond \overline{C}$ each word $\alpha \in (U \cup O)^\star$ such that the sets $\alpha_{\Downarrow U}$ and $A_{\Downarrow U}$ are disjoint (these words cannot occur in the composition). Given the equation $A \diamond X \subseteq C$ over prefix-closed regular languages accepted by a finite automaton with nonaccepting states, a procedure leading to the same incomplete solution $S = A \diamond C \setminus A \diamond \overline{C}$ is presented and argued to be maximal in [55]. Notice that the version of Procedure 2.3.1 for parallel composition handles unrestricted regular languages, so it is more general than the ones in [94] or [55]. Safe and compositionally deadlock-free solutions are considered in [36] and a procedure to derive the largest safe solution is proposed.

In [107], the equation $A \diamond X \subseteq C$ is solved over complete FSMs. It is shown that the solution over LTSs cannot be applied directly to FSMs because in general it is not an FSM language (an FSM language must be also prefix-closed and I-progressive). Moreover, compositionally $(U \cup O)$-convergent solutions are considered. A technique is proposed to find the so-called largest candidate solution by deleting from the largest solution the sequences causing deadlocks in the composition with FSM A. If the obtained solution is compositionally $(U \cup O)$-convergent then it is called the largest solution of the equation; otherwise, the question whether a compositionally $(U \cup O)$-convergent solution exists remains open. In the latter case the authors propose to limit to l the number of internal interactions; if, for a given l, the largest solution of the equation exists then it is compositionally $(U \cup O)$-convergent and also each of its reductions inherits this property. Notice that in Sect. 3.4.3 we extended to regular languages the procedure for determining the solution with a limited number of internal interactions.

A number of papers in process algebra [36, 55, 94, 107, 118] discuss the equation $A \diamond X \approx C$ under various relations $\approx$, where A and C are given processes.

5.2.2 *Supervisory Control*

In supervisory control a controller (or supervisor) restricts the behavior of a plant by dynamically disabling some of the controllable events after the execution of each event with the goal to match a desired behavior. An extended literature has been developed since the seminal work by Ramadge and Wonham, studying control under complete and partial observation, centralized and distributed control and also control of non-terminating behaviors [76, 77, 101, 119].

Restricted solutions have been investigated too, e.g., in [78] the authors proposed a method to derive the largest compositionally progressive solution over finite automata. The method is based on splitting the states of the automaton of the unrestricted solution, combining the new automaton with the context and then deleting from the composed automaton the states that are not completely specified (for at least an input there is no transition). The method is proposed for a general supervisory control topology where the language of the specification C is the restriction on the external alphabet of a subset of the language of the known component A. A generalization of these results for a more general topology has been proposed in [21]. Restricted solutions to parallel equations over regular languages are considered also in [76, 101]. In [76] the solution is called the minimally restrictive supervisor, i.e., a supervisor that combined with the context matches the largest specification sublanguage; in [101], a bound l is set on the number of internal interactions. For updates on control of discrete event systems, including fault tolerant control and Petri nets the reader may refer to [61, 64].

The computational complexity of the harder supervisory control problems (verification and synthesis) varies from NP-hard/complete to (most often) PSPACE-hard/complete (see [48, 123, 124]); some proofs use the reduction from the automata intersection problem.

5.2.3 *Simulation Relations*

A procedure for deriving the largest solution for LTSs A and C of the equation $A \diamond X \preceq^{bisim} C$ was proposed in [118], where $\preceq^{bisim}$ denotes either a strong or weak bisimulation relation (as opposed to language containment in our approach). The authors derive an automaton representing all traces of the parallel composition of A and each sequence over the alphabet $U \cup O$. The final state of each trace of the composition whose $I \cup O$-restriction is not a trace of C is marked as a *bad* state. To compute the largest solution, the automaton is augmented with a designated *don't care* state that accepts all infeasible sequences, i.e., those that do not occur in the composition. Finally the $(U \cup O)$-restriction of the automaton is derived and the result is determinized by subset construction, where each subset including a bad state is marked as bad. The largest solution is the subset of all sequences of the $(U \cup O)$-restriction whose runs do not end up in bad states. In [155] the authors study supervisory control for nodeterministic specifications to design a controller such that the closed-loop system is bisimilar to the specification. Nondeterministic plant and specification are studied also in [59], using the trajectory model.

5.2.4 Delay-Insensitive Processes

In the context of modeling delay-insensitive processes and their environments, a number of concurrency models use various labelings of states of processes to represent certain properties of states, such as quiescence and error or violation [90, 100]. The existence of state labels requires a stronger semantics than language semantics and leads to a reflection operator further refining the language complementation operator. See [90, 100] for discussions on parallel composition operators for delay-insensitive processes. The largest solution of the equation $P \parallel X \leq R$ is the process $\tilde{\ }(P \parallel \tilde{\ } R)$, where $\parallel$ is a composition operator and $\tilde{\ }$ is a reflection operator, replacing the complementation operator used with language semantics. As with language semantics, such a solution might not be compositionally $(U \cup O)$-convergent; it is also of interest to look for solutions exhibiting a property called *healthness*, capturing correctness properties according to the chosen parallel composition operator [90].

5.3 ω-Automata, Games, Realizability and Synthesis

Given a game between two players, player-0 and player-1, in a single step of the game player-0 chooses a symbol from the finite alphabet Σ_0 and player-1 does the same from the finite alphabet Σ_1. A play of the game is a finite or infinite sequence of steps. Player-0 wins the game if the play is in the game language, $\Gamma \subseteq (\Sigma_0 \times \Sigma_1)^\omega$; otherwise, player-1 wins, i.e., when the play is in $\overline{\Gamma}$.

A strategy for a player is a rule that tells the player what symbol to choose at each instant; a winning strategy is one that insures a win for the player no matter how the other player behaves. Each player sees the move of the other player at each step, and the strategy may be a function of the entire history of the game, a game of perfect information. The game is determined if one of the players has a winning strategy.

This type of infinite games of perfect information were first considered by Gale and Stewart [45] (GSP games). If Γ is presented as an ω-automaton. we have a Gale-Stewart game of perfect information on an ω-automaton. In this case the winner has a finite-state strategy implemented by an FSM. Büchi and Landweber [20] gave a constructive procedure to solve games on ω-automata, i.e., to decide the winner and synthesize the FSM representing the winner's strategy. Solving a game on an ω-automaton is also known as Church's problem [31]. If Γ is a property to be satisfied by a synchronous digital circuit to be synthesized, what we need to know is whether the circuit to be synthesized has a winning strategy against its adversarial environment (problem of *realizability*), and if it exists, we need to find an FSM realizing a winning strategy subject to the implementation constraints (problem of *synthesis*).

Notice that while for combinational logic synthesis, given a Boolean relation $R \subseteq \{0,1\}^n \times \{0,1\}^m$, there is always a Boolean function $f : \{0,1\}^n \to \{0,1\}^m$ satisfying the relation R, i.e., such that for every $i \in \{0,1\}^n$, if there exists $o \in \{0,1\}^m$ for which $(i,o) \in R$, then $(i, f(i)) \in R$ [17], for sequential logic synthesis this is not always the case. Indeed, the objective of sequential synthesis is to come up with a sequential circuit that satisfies a property relation $R \in I^\omega \times O^\omega$ specifying for each input sequence the expected output sequences. Such a circuit can be seen as an FSM whose behavior corresponds to a function $f : I^\star \to O$ such that, for all $i_0 i_1 i_2 \cdots \in I^\omega$ and $f(i_0)f(i_0 i_1)f(i_0 i_1 i_2)\cdots \in O^\omega$, it holds that $(i_0 i_1 i_2 \ldots, f(i_0))f(i_0 i_1)f(i_0 i_1 i_2)\cdots \in R$. Not every property is realizable, i.e., there is not always an FSM that implements it; e.g., the relation $R \subseteq \{0,1\}^\omega \times \{a,b\}^\omega$ defined by $(01\{0,1\}^\omega, a^\omega) \in R$ and $(01\{0,1\}^\omega, b^\omega) \in R$ is not realizable [74], because after the first input 0 to decide whether to output a or b it would need to look into the future to know whether the second input is 1 (and then output a now) or 0 (and then output b now).

However property synthesis problems arising in practice, such as the synthesis of the unknown component problem, do no always fall in the paradigm of Gale-Stewart games of perfect information. To model these synthesis problems, Krishnan and Brayton [74] proposed and studied the paradigm of two-person Gale-Stewart games of incomplete information on deterministic ω-automata (GSI games). For these games each player observes only the observable part of the opponent's actions, on which he has to base his moves. The winner of GSI games can be decided and a winning strategy synthesized by reducing them to pairs of appropriate GSP games, at a higher computational cost. Some GSI games are not determined, i.e., there is no winner with a deterministic winning strategy.

More precisely, GSI games are two person games played between two players: player-0 and player-1. In a single step of the game player-0 chooses a symbol from the finite alphabet $\Sigma_0 = \Sigma_0^{obs} \times \Sigma_0^{hid}$, followed by player-1 choosing a letter from the finite alphabet $\Sigma_1 = \Sigma_1^{obs} \times \Sigma_1^{hid}$. Same as for GSP games, a play of the game is a finite or infinite sequence of steps. The distinguishing feature of GSI games (Gale-Stewart games of incomplete information or partial observation) with respect to GSP is the fact that each symbol in the players' alphabet has two parts: the observed (Σ^{obs}) and the hidden one (Σ^{hid}). A player only sees the observed part of the opponent's play: e.g., if player-0 selects $\sigma_0 = (\sigma_0^{obs}, \sigma_0^{hid})$, player-1 can only see σ_0^{obs}; similarly, player-0 observes only the Σ_1^{obs} part of player-1's moves. Each player instead knows both the observed and hidden part of the own move.

An interpretation of the synthesis of the unknown component as solving a GSI game between a context player-0 M (with language $L(M) \subseteq (I \times V \times U \times O)^{\omega}$), an unknown player-1 C (with language $L(C) \subseteq (V \times U)^{\omega}$), and a specification $S \subseteq (I \times O)^{\omega}$ has been offered in [74]. It is shown that the set of winning strategies for player-1 is the set of solutions for the Watanabe-Brayton rectification problem. A similar statement has been made for the controller's topology.

We are indebted for the exposition in this Section to the Ph.D. Dissertation by Krishnan [74].

Problems

5.1. Hierarchical Optimization vs. Global Optimization
Rho and Somenzi observed in [121] that there are cases when M_A is optimal with respect to M_B and vice-versa (i.e., there is no flexibility in the series composition of M_A with respect to M_B nor vice versa of M_B with respect to M_A), yet their cascade interconnection (M_A feeding M_B) is not the smallest implementation of the product FSM. To prove their point they exhibited two FSMs M_A and M_B reported respectively in Figs. 5.5 and 5.6.

(a) Minimize the states of M_A and M_B in isolation.
(b) Are there input sequential don't cares of M_A to optimize M_B ?
(c) Are there output sequential don't cares of M_B to optimize M_A ?
(d) Merge the two FSMs M_A and M_B and minimize the collapsed FSM so obtained.
(e) Comment the previous results. What are the features of these two machines that explain their behavior in composition ?

5.2. Optimization of a Head Component in a Series Topology
Figure 5.7 shows a series topology $M_A \rightarrow M_B$, with no input don't care sequences of M_B (since the output sequences of M_A coincide with $\{0, 1\}^{\star}$), where M_B is

Input	Present state	Next state	Output
0	A	B	1
1	A	C	0
0	B	A	0
1	B	D	1
0	C	D	0
1	C	B	1
0	D	C	1
1	D	B	0

Fig. 5.5 Machine M_A for Problem 5.1

Fig. 5.6 Machine M_B for Problem 5.1

Input	Present state	Next state	Output
0	a	c	1
1	a	b	0
0	b	a	0
1	b	d	1
0	c	d	0
1	c	b	1
0	d	b	1
1	d	c	0

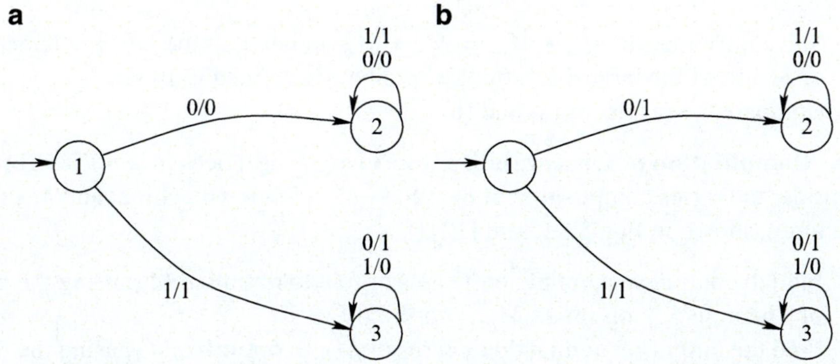

Fig. 5.7 Components (**a**) M_A and (**b**) M_B of series topology $M_A \to M_B$ of Problem 5.2

information-lossless and state-reduced, and so Devadas' and Rho's procedures find no sequential output don't cares for M_A. Let the output value u_i be associated to the edge e_i of M_A, so that $(u_1, u_2, u_3, u_4, u_5, u_6)$ denotes an assignment of output values to M_A. E.g., in Fig. 5.7a it is: $e_1 = 1 \xrightarrow{0/0} 2, u_1 = 0$; $e_2 = 2 \xrightarrow{1/1} 2, u_2 = 1$; $e_3 = 2 \xrightarrow{0/0} 2, u_3 = 0$; $e_4 = 1 \xrightarrow{1/1} 3, u_4 = 1$; $e_5 = 3 \xrightarrow{1/0} 3, u_5 = 0$; $e_6 = 3 \xrightarrow{0/1} 3, u_6 = 1$. Then one can verify that each of the following assignments preserves the behavior $M_A \to M_B$:

$$(u_1, u_2, u_3, u_4, u_5, u_6) = \begin{cases} (0, 1, 0, 1, 0, 1) \\ (0, 1, 0, 0, 1, 0) \\ (1, 0, 1, 0, 1, 0) \\ (1, 0, 1, 1, 0, 1) \end{cases}$$

Notice that the states reachable in M_A change according to the output function of M_A.

(a) Find out what is *best* assignment with respect to the final implementation of M_A measured by the number of states and/or its logic cost.

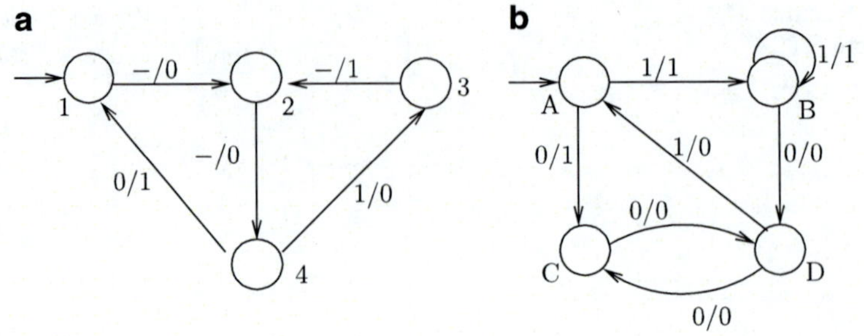

Fig. 5.8 Example for Problem 5.2; **(a)** Driving machine M_A; **(b)** Driven machine M_B

(b) Solve the equation $M_X \bullet M_B = M_A \bullet M_B$ and extract the *best* implementation of M out of the largest solution, where *best* is defined as in (a).
(c) Compare the results of (a) and (b).

5.3. Computation of Observability Don't care Sequences in a Series Topology
Consider the series composition of the FSMs M_A (driving machine) and M_B (driven machine), shown in Fig. 5.8 (from [121]).

(a) Find the output sequential don't cares of M_B to optimize M_A, using the method of Theorem 5.3; optimize M_A accordingly.
(b) Find the output sequential don't cares of M_B to optimize M_A, using the method of Theorem 5.4; optimize M_A accordingly.
(c) Find the largest flexibility for M_A solving the equation $M_X \bullet M_B \subseteq M_A \bullet M_B$; optimize M_A accordingly.
(d) Compare the results of the different techniques.

5.4. Games on ω-automata
In this exercise we follow the exposition in [74]). The game is played on a deterministic ω-automaton $\mathcal{G} = (Q, \Sigma_0 \times \Sigma_1, \Delta, q_0, \phi)$, called the game automaton. The game starts at state q_0; player-0 plays a $\sigma_0^0 \in \Sigma_0$ and then player-1 plays a $\sigma_1^0 \in \Sigma_1$, and the game advances to state $q_1 = \delta(q_0, (\sigma_0^0, \sigma_1^0))$, player-0 then plays a $\sigma_0^1 \in \Sigma_0$, followed by player-1 playing a $\sigma_1^1 \in \Sigma_1$ and so on forever. The infinite sequence $\sigma = (\sigma_0^0, \sigma_1^0)(\sigma_0^1, \sigma_1^1)\ldots$ is a play of the game. The acceptance condition ϕ of the game automaton is the winning condition for player-0 and $\neg\phi$ is the winning condition for player-1. The language of the game automaton, $L(\mathcal{G})$ is the game language. A play $\sigma \in (\Sigma_0 \times \Sigma_1)^\omega$ is a win for player-0 if $\sigma \in L(\mathcal{G})$, i.e., the set of states visited infinitely often by σ satisfies the winning condition. Otherwise, the play is a win for player-1, whose winning condition is $\neg\phi$. In a Büchi game, player-1's winning condition is a co-Büchi winning condition.

Consider the Büchi game shown in Fig. 5.9a (from [74]).

(a) Show a win play for player 1 and a win play for player-0.
Solution.
The play $(b, 1)(b, 1)(a, 0)^\omega$ is a win for player-1, whereas the play $(b, 1)^\omega$ is a win for player-0.

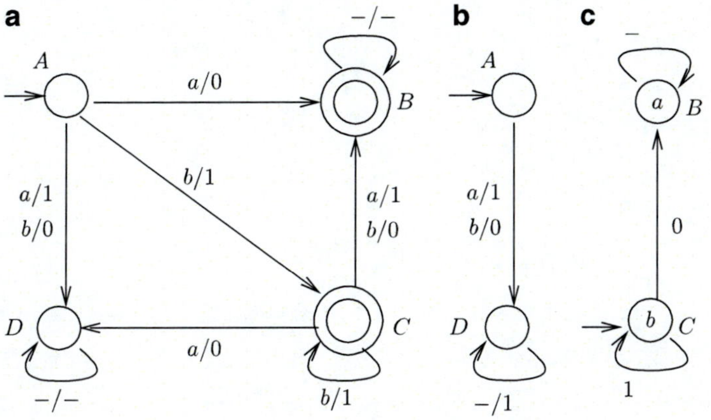

Fig. 5.9 (**a**) Büchi game; (**b**) Winning strategy for player-1 starting as second from state A; (**c**) Winning strategy for player-0 starting as first from state C

(b) Show a winning strategy for player-1 starting as second from state A.
Solution.
See Fig. 5.9b.

(c) Show a winning strategy for player-0 starting as first from state C.
Solution.
See Fig. 5.9c (the output at a state is the letter written inside the state).

Part II
Algorithms for Solving FSM Equations: BALM

The first part of this book presented the theory of language equations for synchronous composition. The theory was developed with several objectives in mind:

- Generalize and unify several known approaches to solve the unknown component problem within the framework of solving language equations.
- Provide a uniform method for expressing several classes of practical problems.
- Enable an efficient software implementation to explore the limits of practical applications.
- Provide a platform on which new algorithms can be developed and tested.

In Part II, we focus on the efficient implementation aspects of this work. A software system was developed, BALM, the Berkeley Automata and Language Manipulation system, based on the multi-valued logic synthesis environment, MVSIS [46]. Both software packages and their documentation are available from the Download Software link on the web site of the UC Berkeley MVSIS: Logic Synthesis and Verification Group [52]. When implementing the theory, we had the following objectives:

- To validate the theory as well as to refine it during its development when counter-examples showed that some conjectures did not hold.
- To solve some practical problems by making the implementation as efficient as possible.
- To facilitate instruction about finite automata, by developing a generic finite automata calculator supported by the ability to graphically display example results.

BALM utilizes the programming environment of MVSIS as much as possible.[1] The implementation was organized as a set of modules to facilitate porting, reuse in other applications, and ease of maintenance and extension. Multiple implementations of

[1] An environment with multi-valued variables was chosen because such variables enable easier specification of automata or FSMs and, at the same time, facilitate debugging.

some of the basic algorithms allow comparative studies. BALM is intended to be used in two modes:

- As a black box, where a user invokes a synthesis script (an automatic sequence of high-level commands to BALM) to solve a given problem. In-depth knowledge of the underlying implementation details is not required.
- As a white box, where BALM is used as a platform that can be programmed for other applications dealing with finite automata and FSMs. Detailed knowledge about programming with MVSIS/BALM is required.

In Chap. 6, we describe some aspects of the implementation of BALM to aid the second type of user. Chap. 7 details how a particularly efficient implementation can be done if (a) the languages are represented initially as deterministic sequential circuits, and (b) the solution is required as an FSM language. Chap. 8 is a summary of how BALM can be used as a black box for a user who wants to try BALM on examples. Chap. 9 illustrates how BALM can be used as an educational tool to show the results of some of the basic automata operations.

Chapter 6
Implementation of Automata Manipulations

In the following, we use the term transition to refer to a directed arc of an STG, which enables the movement of an automaton from the current state to the next state. The transition predicate indicates when the transition is enabled. A transition predicate is an arbitrary Boolean function in terms of the inputs. In some cases, this function is just a minterm. In this case, to avoid confusion, we will refer to a minterm transition.

A number of practical applications require representation and manipulation of finite automata and FSMs. Depending on an application, these objects can be specified as:

1. Sequential logic circuits
2. Transition relations
3. Explicit state transition graphs (STGs)

In the first case, logic circuits are typically represented as directed acyclic graphs with nodes corresponding to logic gates or latches. In the second case, the transition relations are often represented using binary decision diagrams (BDDs) in a monolithic or partitioned form (conjunction of transition relations). A transition relation for an automaton is the set of (current state, next state, predicate) tuples describing the arcs of the automaton. In the third case, an STG specifying an automaton or FSM is represented using one of the three methods described below. The implementation of several procedures for manipulating STGs is outlined in the following subsections.

Languages used in solving language equations are represented as finite automata. Operations on languages are implemented as operations over finite automata. Basic procedures for efficient manipulation of automata assume their specification as STGs. This is the reason that representing and manipulating STGs is of particular importance for implementing the theory developed in this book.

First, we introduce several approaches to representing STGs in a software implementation and motivate the use of one of them, called the hybrid representation, which we believe is the most efficient in general.

T. Villa et al., *The Unknown Component Problem: Theory and Applications*,
DOI 10.1007/978-0-387-68759-9_6,

- *Explicit methods* represent STGs in such a way that every state, transition, and predicate is stored in the computer memory as a separate object.
- *Implicit methods* represent STGs as transition relations in a monolithic or partitioned form. A monolithic form uses one relation in contrast to a partitioned form which has many relations, where it is implied that they are to be conjoined to form a single relation. Transition relations are expressed and manipulated using Binary Decision Diagrams (BDDs). The choice of BDDs is motivated by the fact that they lead to efficient quantification of variables while other representation often have problems with quantification.
- *Hybrid methods* use a combination of explicit and implicit representations.

When implicit methods are used, states and transitions are not represented as specific objects. Instead the BDD is treated as a single object encoding all states and transitions. For many practical problems, the implicit representation is smaller than the explicit one. Implicit representations are often preferred in computations, such as reachability analysis. However, for several important classes of applications, and for most large practical functions, the size of implicit representations can exceed the amount of available memory, which makes implicit representations impractical.

While implementing the proposed methods, we experimented with several types of STG representations. For example, we implemented determinization by subset construction using three methods, explicit, implicit, and hybrid, and compared their performance. The hybrid method was the winner in the majority of cases. The performance of this method scaled well with the problem size.

The above observation motivated the use of a hybrid representation in our software implementation of language solving. In the subsections below, we focus on the hybrid representation, in which, states and transitions are stored explicitly while the predicates are stored implicitly as BDDs. A similar conclusion about the applicability of a hybrid representation to manipulating automata and FSMs was reached in the previous work [139].

6.1 Hybrid Representation of Finite Automata

In this subsection, we describe the hybrid representation of finite automata. When this representation is used, the STG is represented as a graph with explicitly listed transition predicates. This is similar to the explicit representation. The difference is that the predicate enabling a transition from one state to another is represented using a BDD rather than a sum-of-product (SOP).

The reason why the STG is represented explicitly (as a list of nodes and predicates) is that, in most practical problems, the STG is given in this form as an input to the program, or it is derived by enumerating states reachable from an initial state. In both cases, each state is visited at least once during file reading or during STG extraction from network form. As a result, there is no need to develop an implicit representation; the states can just as well be stored explicitly and visited in some order.

The reason why an implicit BDD-based representation is used for the transition predicates is that

- BDDs are efficient for functions with small support.
- BDDs allow for sharing identical or related functions and reusing the computed results.
- BDDs are canonical, which allows for the use of BDD node pointers as unique keys in a hash table.

These advantages are not available when SOPs are used to represent the predicates.

The proposed hybrid representation was tested on a number of practical problems. It was found useful for problems with up to 100,000 states and 25 inputs/outputs. The number of states and transitions can be increased at the cost of linear increase in memory and computation time. Increasing the number of variables above 25 is also possible. However, in this case, the non-robustness of BDD representation limits the scalability because they can blow up in size unpredictably.

6.2 Implementation of Particular Commands

In this subsection, we present implementation details of the procedures performing finite automata manipulations, which are used during language equations solving.

6.2.1 STG Extraction

In many practical applications, an FSM is given as a binary or multi-valued sequential network, for which the initial state is known. The FSMs STG implied by this network can be extracted by an algorithm presented below. The resulting STG is used in a variety of computations.

The *extracted STG* consists of a set of nodes which represent the set of states of the FSM reachable from the initial state. Each state is associated with a set of arcs originating in this state. Each arc has the next state, the output produced, and an input/output predicate, which enables this transition. In the application dealing with finite automata, inputs and outputs are treated uniformly. Therefore, in the STG extraction procedure, we do not distinguish inputs and outputs and simply record transitions as relations depending on both.

A self-explanatory pseudo-code of the STG extraction algorithm is shown in Fig. 6.1. The input to the algorithm is a sequential network with initial states and a limit on the number of states extracted. The output is the hybrid representation of the resulting STG. The computation terminates if the limit on the number of states has been reached.

```
stgExtract( network, limit ) {
    /* initialize the STG to be extracted */
    STG = newSTG();
    /* get hold of the initial state of the network */
    I = getInitialState( network );
    /* initialize the set of states to be explored with the initial states */
    S = I;
    /* explore states */
    while ( S ≠ ∅ ) {
        /* extract one state from the set of states to be explored */
        s = getOneState( S ); S = S \ s;
        /* get the relation of the given state */
        R(i, o, x) = deriveStateRelation( network, s );
        /* enumerate next states reachable from the given state */
        while ( R(i, o, x) ≠ ∅ ) {
            /* find one combination of variables from the state relation */
            minterm(i, o, x) = getOneCombination( R );
            /* get the next state of this combination */
            n(x) = ∃_io minterm(i, o, x);
            /* get the IO predicate of this next state */
            pred(i, o) = ∃_x minterm(i, o, x);
            /* remove transition to this state from the relation */
            R(i, o, x) = R(i, o, x) . !n(x);
            /* if the state does not exist, schedule it for exploration */
            if ( stateIsNew(STG, n) )
                S = S ∪ n;
            /* record the transition */
            addTransition( STG, s, n, pred );
        }
        /* quit if the number of states exceeded the limit */
        if ( getNumStates(STG) ≥ limit )
            return NULL;
    }
    return STG;
}
```

Fig. 6.1 STG extraction algorithm

The internal procedures called from the STG extraction algorithm are the following:

newSTG: Creates a new state transition graph.
getInitialState: Queries the network for its initial state.
getOneState: Extracts one state from the set.
deriveStateRelation: Traverses the network and derives the transition relation from the given state as a BDD in terms of inputs, i, outputs, o, and next state variables, x.

```
completeAutomaton( A ) {
    /* iterate over states and check if there are incomplete states */
    /* cond(i)_{s→t} is the predicate over input variables i */
    /* of the transition from state s to state t */
    complete = 1;
    for each state s of automaton A {
        sum(i) = 0;
        for each transition t from state s
            sum(i) = sum(i) + cond(i)_{s→t}
        if ( sum(i) ≠ 1 ) {
            complete = 0;
            break;
        }
    }
    if ( complete )
        return;
    /* create the non-accepting state na */
    addState( A, na );
    /* iterate over states and add transitions from
    the incomplete states into the non-accepting state na */
    for each state s of automaton A {
        sum(i) = 0;
        for each transition t from state s
            sum(i) = sum(i) + cond(i)_{s→t}
        if ( sum(i) = 1 )
            continue;
        addTransition( A, s, na, !sum(i) );
    }
}
```

Fig. 6.2 Completion algorithm

getOneCombination: Extracts one minterm (m_i, m_o, m_x) from the BDD.
stateIsNew: Checks if the state m_x is already in the STG.
addTransition: Adds one transition to the STG.
getNumStates: Queries STG for its number of states.

6.2.2 Completion

A finite automaton is complete if the Boolean OR of input predicates of all transitions from each state evaluates to constant 1. If there exists at least one state, for which this predicate does not hold, the automaton is incomplete. The pseudo-code in Fig. 6.2 contains the pseudo-code of the completion algorithm.

The algorithm returns without changing the automaton if it is complete. Otherwise, it creates a new non-accepting state na. Next, the algorithm iterates through the states of the automaton. For each state, the algorithm computes its domain, i.e. the union of all transitions enabled from this state. If it is incomplete, a transition from this state to the non-accepting state is added with the associated transition being the complement of the domain.

6.2.3 Determinization

Determinization is a crucial step in solving language equations and in many other algorithms working with finite automata. Determinization of a finite automaton is required before it is complemented. It derives a deterministic automaton accepting the same language of the original non-deterministic automaton. If the automaton is already deterministic, determinization is not performed.

A self-explanatory pseudo-code of determinization by subset construction (see Sect. 2.3.1) is presented in Fig. 6.3. The input to the procedure is a finite automaton and a limit on the number of states of the determinized automaton. The output is a deterministic automaton that is equivalent to the initial one. The computation will terminate if the limit on the number of states has been reached.

The procedures called from the determinization algorithm are the following:

newAutomaton: Creates a new finite automaton.
getInitialState: Queries the automaton for its initial state.
getOneState: Extracts one subset state from the set.
deriveStateRelation: Iterates through the subset s of states of automaton A, then over the transitions from these states, and derives the transition relation of s as a BDD in terms of inputs, i, and next state variables, x.
getOneCombination: Extracts one arc from the BDD.
stateIsNew: Checks if the end state is already present in the automaton.
addTransition: Adds one transition to the automaton.
getNumStates: Queries automaton for the number of states in it.

A non-trivial part of the above determinization procedure is deriving the subsets of next states, $n(x)$, reachable from the subset s. To this end, a single predicate in terms of variables i and x, *minterm(i, x)*, is found in $R(i, x)$, which is the transition relation of the subset of states, s. Next, the minterm, *input(i)*, in terms of input variables, i, is computed from this minterm, by quantifying state variables, x. The subset of next states, $n(x)$, reachable under minterm *input(i)* is found by computing the image of *input(i)* using the relation $R(i, x)$. The full predicate, *pred(i)*, of the transition from subset s into subset n is computed by universally quantifying next state variables, x, from the expression stating equality of $R(i, x)$ and $n(x)$; we want all those i which precisely go to the subset n and no more or less states. To understand the reason for computing $pred(i) = \forall_x\ [R(i, x) \equiv n(x)]$, suppose as an example that **getOneCombination** selects $minterm(i_1, n_1)$, then it is $input(i_1)$

```
subsetConstruction( A, limit ) {
    /* consider the case of a deterministic automaton */
    if ( automaton A is deterministic )
        return A;
    /* initialize the STG to be extracted */
    B = newAutomaton();
    /* get the initial state of the network */
    I = getInitialState( A );
    /* initialize the set of subset states to be explored with the initial state */
    S = {I};
    /* explore states */
    while ( S ≠ Ø ) {
        /* extract one subset state to be explored */
        s = getOneState( S ); S = S \ s;
        /* derive the transition relation of the given subset state */
        R(i, x) = deriveStateRelation( A, s );
        /* enumerate the next state subsets reachable from the given subset */
        while ( R(i, x) ≠ Ø) {
            /* read one combination of variables in the relation */
            minterm(i, x) = getOneCombination( R );
            /* get the input minterm of this combination */
            input(i) = ∃x minterm(i, x);
            /* get the subset of next states associated with this input */
            n(x) = ∃i [R(i, x) . input(i)];
            /* get the transition predicate associated with this subset */
            pred(i) = ∀x [R(i, x) ≡ n(x)];
            /* remove the predicate to this subset from the relation */
            R(i, x) = R(i, x) . !pred(i);
            /* if the next state subset is new, schedule it for exploration */
            if ( stateIsNew(B, n) ) {
                S = S ∪ n;
                /* set the acceptance attribute of the new product state */
                if ( subset n contains at least one accepting state )
                setAttribute( n, "accepting" );
                else
                setAttribute( n, "non-accepting" );
            }
            /* add transition from state s to state n under predicate pred */
            addTransition( B, s, n, pred );
        }
        /* quit if the number of states exceeds the limit */
        if ( getNumStates(B) ≥ limit )
            return NULL;
    }
    return B;
}
```

Fig. 6.3 Determinization algorithm

and say $n(n_1, n_2, n_3)$ (say that n_1, n_2, n_3 are the next states under input i_1), finally suppose that n_1, n_2, n_3 are exactly the next states also under input i_3, then it will be $pred(i_1, i_3)$.

The transition relation $R(i, x)$ is reduced by removing the transitions to $n(x)$ under predicate *pred(i)*. If state *subset n(x)* is new, it is scheduled for exploration by the proposed algorithm. Finally, transition from s to n under input *pred* is added to the automaton B under construction.

6.2.4 Complementation

Complementation of a deterministic finite automaton is performed by exchanging the set of accepting and non-accepting states. This is done by enumerating all states of the automaton to be complemented and flipping the acceptance attribute of each state.

If the automaton is not deterministic, it must be processed by the determinization algorithm first.

6.2.5 Support

Support is an operation, which resets the support of the automaton to be equal to the given ordered list of variables. The following three situations may occur:

- If a listed variable is not currently in the support, it is added to the resulting support and the domain of all transition predicates is expanded to include this variable. Since the predicates are represented implicitly using BDDs, the manager is expanded to include the new variable, while the BDDs of the predicates do not have to be modified, because they do not depend on the new variable.
- If a variable already exists in the support, it is simply moved to its new position in the given order. This does not affect BDD ordering, so BDDs do not change, however it changes the mapping function that matches the ordered list of support variables and the internal BDD variables.
- If a variable that is currently present in the support is not listed for inclusion in the resulting support, it is removed from the support. This is done by quantifying it existentially from all the predicates enabling the transitions of the given automaton.

In the end, when the new variable order is defined, the variable map of the BDDs representing all predicates is updated while the predicates remain unchanged. The resulting automaton is returned.

6.2.6 Product

Product of two finite automata, A and B, is a new automaton whose states are pairs of states from A and B, respectively, reachable from the initial states of A and B under some input sequences.

It is assumed that both A and B are complete when forming the product. This can be achieved by applying the operation *Completion* to an automaton that is not yet complete: in this way it is completed by adding the non-accepting don't care state, and all missing transitions from any state are directed to this non-accepting don't care state.

It is assumed that both A and B have the same support when forming the product. This can be achieved by forcing them to have the same support, if it is not already the case, by the operation *Support*, which resets the support of an automaton to be equal to a given ordered list of variables. For instance suppose that the support of automaton A is (x, z) and the support of B is (y, z), then the user must set the supports of both A and B to be x, y, z.

If A and B share a variable in their supports, it is identified as the same variable and they must agree in value. Suppose that A is in state s_a and B is in state s_b, and that A transits to accepting state $\tilde{s}_a$ on input $(\alpha, -, \gamma)$ and B transits to accepting state $\tilde{s}_b$ on input $(-, \beta, \gamma)$ (the input to $'-'$ means don't care as an effect of lifting a variable to get a common support), then the product automaton transits from product state (s_a, s_b) to product state $((\tilde{s}_a, \tilde{s}_b)$ under input (α, β, γ). A product state is accepting if and only if both of the component states are accepting.

A self-explanatory pseudo-code of the algorithm computing the product of two automata is presented in Fig. 6.4. The input to the procedure are the automata and the limit on the number of states of the product automaton. The output is the product automaton. The computation will exit if the limit on the number of states is reached.

6.2.7 Prefix-Closed

Prefix-closed is an operation applied to a finite automaton, which transforms it as follows. The non-accepting states are removed together with all the transitions into them and from them. Only the states reachable under this new automaton from the given initial state are kept. This operation returns the empty automaton if the initial state is not accepting.

6.2.8 Input-Progressive

Input-progressive is an operation, applied to a finite automaton, which transforms it as follows. First, operation prefix-closed is applied. If the resulting automaton is not

```
productAutomaton( A, B, limit ) {
    /* initialize the product automaton */
    P = newAutomaton();
    /* get initial states of the automata */
    I_A = getInitialState( A );
    I_B = getInitialState( B );
    /* initialize the set of states to be explored with the initial product state */
    S = {I_A, I_B};
    /* explore states */
    while ( S ≠ Ø ) {
        /* extract one product state from the set of states to be explored */
        {s_A, s_B} = getOneState( S ); S = S \ {s_A, s_B};
        /* iterate through the transitions from these component states */
        for each state t_A reachable from state s_A in automaton A {
            for each state t_B reachable from state s_B in automaton B {
                /* transition conditions must agree on common support */
                cond(i) = cond(i)_{s_A→t_A} · cond(i)_{s_B→t_B};
                if ( cond(i) = Ø )
                    continue;
                // if the new product state does not exist, add it for exploration
                if ( stateIsNew(P, {t_A, t_B}) ) {
                    S = S ∪ {t_A, t_B};
                    /* set the acceptance attribute of the new product state */
                    if ( stateIsAccepting(t_A) and stateIsAccepting(t_B) )
                        setAttribute( {t_A, t_B}, "accepting" );
                    else
                        setAttribute( {t_A, t_B}, "non-accepting" );
                }
                /* record the transition */
                addTransition( P, {s_A, s_B}, {t_A, t_B}, cond(i) );
            }
        }
        /* quit if the number of states exceeded the limit */
        if ( getNumStates(P) ≥ limit )
            return NULL;
    }
    return P;
}
```

Fig. 6.4 Product computation algorithm

empty, it is processed iteratively. Each iteration removes any state whose outgoing transitions under some input are undefined. The iteration stops when the automaton is not changed. This operation may return the empty automaton. Note that the subset of the support variables that constitute the inputs must be specified as an input to this procedure.

Problems

6.1. Show that the subset construction to convert a non-deterministic finite automaton into a deterministic finite automaton can be performed in $O(kn2^n)$ time, where k is the alphabet size and n is the number of states.

6.2. State Minimization
Algorithms of different complexity and ease of implementation for state minimization of deterministic automata have been reported in the literature. For this problem we rely on the excellent discussion available in [126], where four algorithms are mentioned: **naive-minimize**, **minimize**, **fast-minimize** and **brzozowski**.

(a) **naive-minimize** has worst-case complexity $O(n^3)$. The following pseudo-code is proposed in [126], p. 83:

```
naive-minimize(Aut) {
    for (all unordered pairs (p, q), p ≠ q)
        U[(p, q)] := 0;
    for (all unordered pairs (p, q), p ∈ F, q ∈ Q \ F)
        U[(p, q)] := 1;
    done := false;
    while (not(done)) {
        done := true;
        T:= U;
        for (each unordered pair (p, q), with T[(p, q)] = 0)
            for (each a ∈ Σ)
                if (T[(δ(p, a), δ(q, a))] = 1) {
                    U[(p, q)] := 1;
                    done := false;
                }
    }
    return(U)
}
```

(b) **minimize** has worst-case complexity $O(n^2)$. The following pseudo-code is proposed in [126], p. 87:

```
minimize(Aut) {
    for (all unordered pairs (p, q), p ≠ q)
        U[(p, q)] := 0;
    for (all unordered pairs (p, q), p ∈ F, q ∈ Q \ F)
        U[(p, q)] := 1;
    for (each unordered pair (p, q), with either p, q ∈ F or p, q ∉ F)
        if (U[(δ(p, a), δ(q, a))] = 1 for some symbol a ∈ Σ) {
            U[(p, q)] := 1;
            recursively set U[(p', q')] := 1 for all unmarked pairs (p', q') on
the
            list for (p, q), and all pairs on those lists, etc.;
```

```
            }
            else {
                for (all a ∈ Σ)
                    if(δ(p,a)≠δ(q,a)) put (p,q) on the list for (δ(p,a), δ(q,a));
            }
        return(U);
    }
```

(c) **fast-minimize** has worst-case complexity $O(n \log n)$. Due to Hopcroft [14,62], the best exposition is by Gries [51].

(d) **brzozowski** has worst-case complexity $O(kn2^{2n})$ (but is practice is classified as often good) and it uses a series of four operations: reverse, subset construction, reverse, subset construction. Brzozowski described first the algorithm in 1962 and it has been rediscovered more than once since then, see [18] for the whole story.

Consider the automaton $A = \langle S_A, \Sigma_A, \Delta_A, r_A, Q_A \rangle$, defined as follows: $S_A = \{s0, s1, s2\}$, $\Sigma_A = \{0, 1\}$, $\Delta_A = \{(1,\ s0,\ s0), (0,\ s0,\ s1), (1,\ s1,\ s2), (0,\ s1,\ s1), (1,\ s2,\ s1), (0,\ s2,\ s2)$, $Q_A = \{s1, s2\}$, $r_A = s0$.

Minimize the automaton A with algorithms (a) and (b).

Retrieve from the literature algorithms (c) and (d) and minimize the automaton A with them.

6.3. Show that if A_1 and A_2 are minimal deterministic finite automata recognizing a language L , then A_1 and A_2 are isomorphic, i.e., their graphs are identical up to renaming of states.

Give a counter-example to the previous statement for non-deterministic finite automata.

Chapter 7
Manipulations of FSMs Represented as Sequential Circuits

In many cases, hard computational problems can be reformulated using decomposition and partitioning, which can lead to computational advantages. An example is image computation, which is a core computation in formal verification. In its simplest form, the image of a set of states is computed using the formula:

$$Img(ns) = \exists_{i,cs} \, [T(i, cs, ns) \, . \, \xi(cs)],$$

where $T(i, cs, ns)$ is the transition relation, $\xi(cs)$ is the set of current states, i is the set of input variables, and $cs(ns)$ is the set of current (next) state variables. The image, $Img(ns)$, is the set of states reachable in one transition under all possible inputs from the current states, $\xi(cs)$, using the state transition structure given by $T(i, cs, ns)$.[1]

When the state transition structure is given by a sequential network, the transition relation can be represented, for example, in partitioned form using a set of next state functions, $\{T_k(i, cs)\}$, which update the state of each latch.[2] The partition, $\{T_k(i, cs, ns_k)\}$, represents how the next state of each latch, ns_k, depends on the values of inputs, i, and the current state, cs, of all latches:

$$T_k(i, cs, ns_k) = [ns_k \equiv T_k(i, cs)].$$

The complete transition relations can be derived as the product of the next state functions as follows:

$$T(i, cs, ns) = \prod_k T_k(i, cs, ns_k) = \prod_k [ns_k \equiv T_k(i, cs)].$$

[1] We do not distinguish between sets, relations, and their characteristic functions. Therefore, it is possible to think of $\xi(cs)$, $Img(ns)$, and $T(i, cs, ns)$ as completely specified Boolean functions represented using e.g. BDDs.

[2] In this section, we will illustrate the concepts using next state functions, but the computations also work for the case of non-deterministic relations.

T. Villa et al., *The Unknown Component Problem: Theory and Applications*, DOI 10.1007/978-0-387-68759-9_7, © Springer Science+Business Media, LLC 2012

We call $T(i, cs, ns)$ the *monolithic* representation of the transition relation and contrast it with the representation in terms of the partition, $\{T_k(i, cs, ns_k)\}$. Many problems can be solved using a partitioned representation without ever resorting to the monolithic one, which may be impossible for larger problems. For example, the image computation can be performed using the partitioned representation by scheduling those cs variables, which do not appear in some parts, to be quantified earlier . In practice, this approach to image computation leads to significantly smaller intermediate BDDs, and has been responsible for dramatic increases, both in efficiency and in the sizes of problems that formal verification can solve.

In this chapter, we discuss how to use the partitioned representation for solving language equations of the type $F \bullet X = S$, for the case when the automata for F and S are prefix-closed and represented by multi-level sequential networks. We show how the partitioned representation can be used to perform the basic operations of language equation solving: completing, complementing, determinizing, hiding, and computing the product of finite automata. An important aspect is that we show how most of these operations can be reformulated in terms of image computations, and hence can take advantage of the efficient developments in this area.

7.1 Problem Statement

The interaction of F and X within the specification S is shown in Fig. 7.1.[3] The external input and output variables are represented by symbols i and o, respectively. The internal variables that are the inputs and outputs of X are represented by symbols u and v, respectively.

We assume that the behaviors of the components F and S are represented by multi-level sequential networks, similar to the one in Fig. 7.2. The external variables

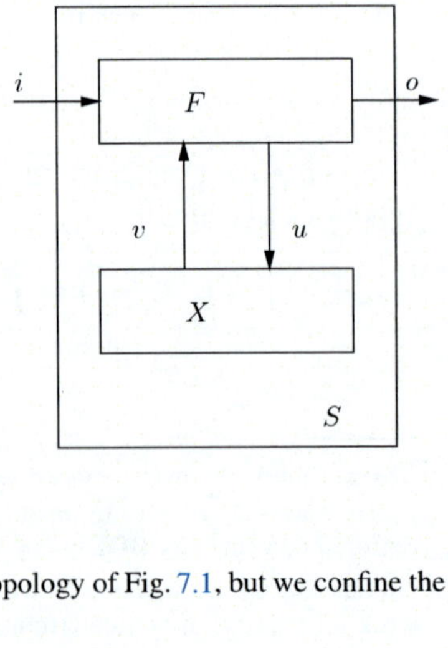

Fig. 7.1 Topology with known component F, unknown component X and specification S

[3]The results of this chapter are not limited to the particular topology of Fig. 7.1, but we confine the discussion to that of Fig. 7.1 for ease of presentation.

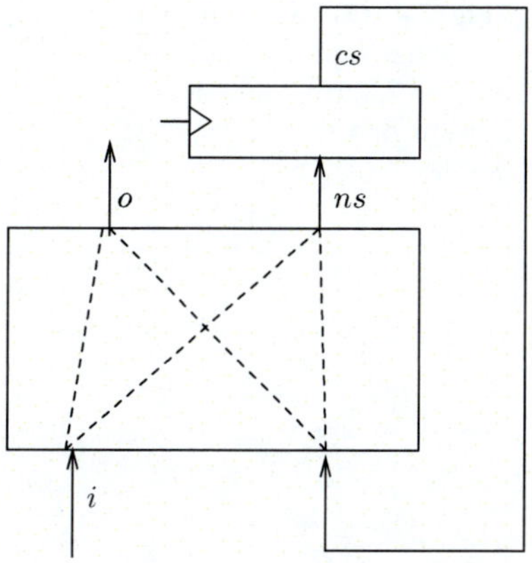

Fig. 7.2 The structure of a sequential network

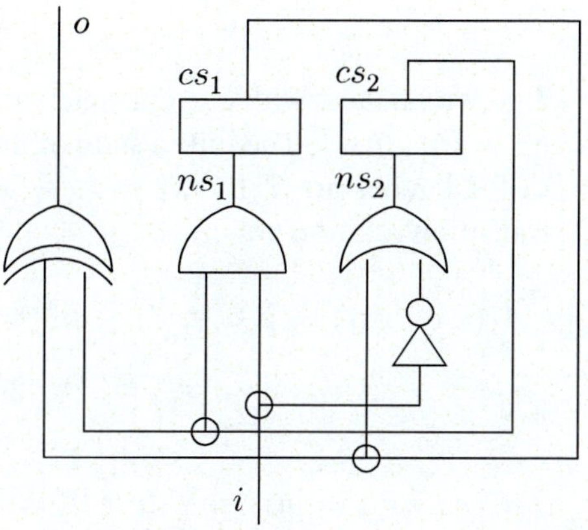

Fig. 7.3 An example of sequential network

of such networks are the primary inputs (i), the primary outputs (o), and the latches having the current state (cs) and next state (ns) variables. We assume that the latch next-state functions, $T_k(i, cs)$, and the primary-output functions, $O_j(i, cs)$, can be computed and stored as BDDs in terms of the primary inputs and the current state variables.

The automata for F and S are derived from the multi-level networks representing them, simply by taking the set of inputs of these automata as the union of the sets of inputs and outputs of the corresponding network. All reachable states of a network are the accepting states of the corresponding automaton, since F and S are FSMs and hence are prefix-closed. A "don't-care" (DC) state can be added to make an automaton complete. This is done by making all automaton input combinations, for which the behavior of a state is not defined, transition to DC, which becomes the only non-accepting state. The requirement of prefix-closed dictates that DC has a universal self-loop.

Example 7.1. Figure 7.3 shows a sequential circuit, while Fig. 7.4a,b show respectively the corresponding FSM and automaton. The circuit has input i, output o,

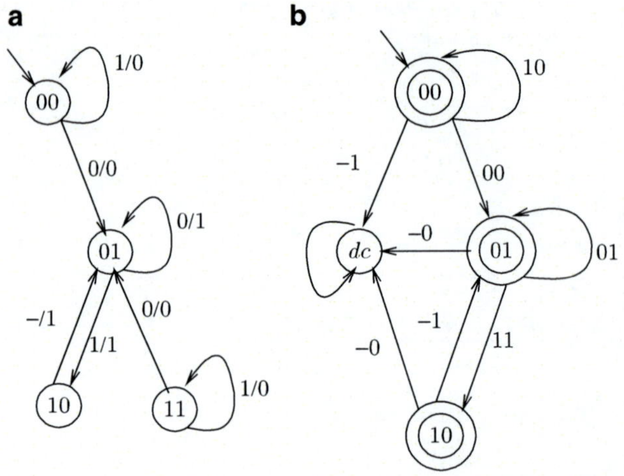

Fig. 7.4 (**a**) FSM extracted from the sequential network of Fig. 7.3; (**b**) Automaton that recognizes the language of the FSM

and two latches, with current state variables $cs = \{cs_1, cs_2\}$ and next state variables $ns = \{ns_1, ns_2\}$. The initial state of the latches is state (00). The next state functions of the latches are $T_1(i, cs) = i.cs_2$ and $T_2(i, cs) = \bar{i} + cs_1$. The two parts of the transition relation are

$$T_1(i, cs, ns_1) = [ns_1 \equiv T_1(i, cs)] = [ns1 \equiv i.cs_2] = ns_1.i.cs_2 + \overline{ns_1}.(\bar{i} + \overline{cs_2})$$

and

$$T_2(i, cs, ns_2) = [ns_2 \equiv T_2(i, cs)] = [ns_2 \equiv (\bar{i} + cs_1)] = ns_2.(\bar{i} + cs_1) + \overline{ns_2}.i.\overline{cs_1}).$$

The single output relation is

$$O(cs_1, cs_2, o) = [o \equiv (cs_1 \oplus cs_2)].$$

The states of the automaton in Fig. 7.4b are labeled with the latch values (cs_1, cs_2). Transitions (arcs) are labeled with (i, o) values. Thus, the transition from state (00) under input 0 is to state (01). The output produced by the network in this case is 0. So the label of this transition is 00 (equal to 0/0 using the conventional FSM labeling, as in Fig. 7.4a). This automaton is not complete. Thus, the transition from (00) under input (11) is not defined. The double-circled states of the automaton are accepting. The additional (single-circled) state (DC), added for completion, is not accepting. DC has a universal self-loop and all transitions that were originally undefined (e.g., from (00) under input (11)) are directed to DC.

7.1.1 Monolithic Representation of Relations

The monolithic representations of the transition and output relations of the automaton can be obtained from the next-state and output functions:

$$T(i, cs, ns) = \Pi_k[ns_k \equiv T_k(i, cs)],$$
$$O(i, o, cs) = \Pi_k[o_k \equiv O_k(i, cs)].$$

The monolithic representation of the complete transition-output relation of the automaton is:

$$TO(i, o, cs, ns) = T(i, cs, ns).O(i, o, cs).$$

TO specifies what the next state is for each current state and each input/output combination. A relation is not well-defined if there exists an input/output/current-state combination for which the behavior of the automaton is not specified, i.e., the automaton is incomplete. As an example of the use of the monolithic representation TO, consider the operation of completion of an incomplete automaton. The transition-output relation of the completed automaton is

$$TO'(i, o, cs, ns) = TO(i, o, cs, ns) + A(i, o, cs).DC(ns) + DC(cs).DC(ns)$$

where

$$A(i, o, cs) = \overline{\exists_{ns}\, TO(i, o, cs, ns)}$$

is the sub-domain of input/output/current-state variables for which the original automaton is not defined, and DC is the code (in terms of latch variables) of the "don't-care" state added during completion. Note that we cannot use the code of an unreachable state to represent the DC because the unreachable states have next states. To encode DC we need to add an additional state variable. Thus the resulting relation is derived from the original one by simply directing all undefined transitions to DC and then adding a universal self-loop to DC.

7.1.2 Partitioned Representation of Relations

The disadvantage of monolithic representations is that they include all variables and hence their BDDs may be huge. Even if these can be computed, the operations, such as completion, product, determinization, become very inefficient, if not impossible, to complete. If the set of reachable states is much smaller than the set of all states, re-encoding the monolithic relations using fewer state bits may alleviate this problem. However, re-encoding can be very slow and, in our experience, this tends to increase the BDD sizes of the relations. We will show how partitioned representations can be used, both in making computations more efficient and in allowing larger problems to be solved. The monolithic transition and output relations will never be constructed. With a partitioned representation, the functionalities of the automata are represented as sets of functions, $\{T_k(i,cs)\}$ and $\{O_k(i,cs)\}$. All Boolean operations are performed using these functions. It will be shown that image computation plays the key role in the manipulations involving partitioned representations, and hence a decade of research resulting in techniques to speed up image computations [30, 120, 137] can be used.[4]

7.2 Computation Algorithms

This section describes how to perform operations on the automata, required for language equation solving, when the partitioned representations are available. The main algorithm for the computation, using either monolithic or partitioned representations, was described in Chap. 3, Sect. 3.3. In this section, we show that, given the partitioned representations, all the steps are essentially embedded into a modified determinization procedure, so that there is no need to perform the completion and variable hiding operations as separate steps. However for generality, we show first how these operations can be performed independently of determinization.

7.2.1 Elementary Operations Using Partitioned Representations

In the application to the problem where the topology is as shown in Fig. 7.1, we have partitions for F (which has both u and o as outputs):

$$\{T^F_{j_{TF}}(i,v,cs^F,ns_{j_{TF}})\},\{U_{j_U}(i,v,u_{j_U},cs^F)\},\{O^F_{j_{OF}}(i,v,o_{j_{OF}},cs^F)\}$$

and partitions for S:

$$\{T^S_{j_{TS}}(i,cs^S,ns_{j_{TS}})\},\{O^S_{j_{OS}}(i,o_{j_{OS}},cs^S)\}.$$

7.2.1.1 Completion

An automaton derived from an FSM is completed by interpreting the complement of its output relation as the condition for a transition from the current state to

[4]It is beyond the scope of this book to describe all these advancements; we experimented extensively with implementing most of proposed techniques and chose the methods that we observed to be most efficient.

the non-accepting DC state. The complement of the output relation, $\overline{O(i,o,cs)}$, gives, for each current state, cs, all the input/output combinations, for which the automaton's behavior is not defined, since a complete FSM is defined for all its input combinations. When the partitioned representation is used, there is no need to construct the monolithic representation of $O(i,o,cs)$. For any current state or subset of current states, given by its characteristic function $\xi(cs)$, its undefined input/output combinations, $O_\xi(i,o)$, can be found by an image computation followed by complementation:

$$O_\xi(i,o) = \overline{\exists_{cs}\ O(i,o,cs).\xi(cs)}.$$

This image can be efficiently computed using the partitioned representation of the output relation:

$$O_\xi(i,o) = \overline{\exists_{cs}\ \Pi_j[o_j \equiv O(i,o,cs)].\xi(cs)}.$$

7.2.1.2 Complementation (Deterministic Case)

In the general case of non-deterministic automata, determinization (subset construction) is required before complementation. In contrast, a deterministic automaton is easily complemented by interchanging the sets of its accepting and non-accepting states. Thus computing $\overline{S}$, which is deterministic, is easy. When an automaton is derived from a deterministic multi-level network, its set of accepting states is equal to the set of reachable states of the network. The DC state introduced during completion is the only non-accepting state. In this case, complementation is performed by changing the interpretation of the DC state: it becomes the only accepting state, while all other states become non-accepting.

7.2.1.3 Product Computation

The product of two automata is defined when they have the same support. In the partitioned view, having the same support simply means that each function is considered as a function of the full set of variables. When the argument automata are represented in their partitioned forms,

$$\left\{T^F_{j_{TF}}\left(i,v,cs^F,ns_{j_{TF}}\right)\right\},$$

and

$$\left\{T^S_{j_{TS}}\left(i,cs^S,ns_{j_{TS}}\right)\right\},$$

the partitioned representation of the product automaton, is simply the union of the two partitions, i.e.,

$$\{T^F_{j_{T^F}}(i, v, cs^F, ns_{j_{T^F}}), T^S_{j_{T^S}}(i, cs^S, ns_{j_{T^S}})\}.$$

The fact that some variables, for example, u and o, are missing, means that the relations are independent of these variables.

7.2.1.4 Hiding Variables

Hiding variables i and o in the automaton representing the product of F and $\overline{S}$ is an operation which changes the support automaton. In the monolithic form, it is performed by existentially quantifying the variables, not in the support, from the transition-output relation TO of the product automaton:

$$TO'(v, u, cs, ns) = \exists_{i,o}\ [TO(i, v, u, o, cs, ns)].$$

This operation usually leads to a non-deterministic automaton, even if the original automaton is deterministic. Hiding variables cannot be performed on the partitioned representation because the operations of existential quantification and product do not commute. For example, it is not possible to quantify the input variables i from the transition relation by quantifying them independently from each partition:

$$\exists_i\ T(i, cs, ns) = \exists_i\ \Pi_j[ns_j \equiv T_j(i, cs)] \neq \Pi_j\ \exists_i\ [ns_j \equiv T_j(i, cs)].$$

Thus hiding cannot be done by acting on the partitions independently. However, a key observation is that the hiding operation can be built into the next step, the determinization procedure, in such a way that there is no need to derive the monolithic transition relation and then apply hiding to it.

7.2.1.5 Determinization (Subset Construction)

Determinization is performed by enumerating explicitly the subsets of states of a non-deterministic automaton, which are reachable from the initial subset state (which contains only the initial state). The basic step is the computation of the subset of states, reachable under various assignments of (u, v) from a given subset of states, $\zeta(cs)$. Denote by $P_\zeta(u, v, ns)$ the sets of states (the subset states) reachable from $\zeta(cs)$ under various combinations of (u, v). Using the monolithic representation, with variable i hidden, the computation would be:

$$P_\zeta(u, v, ns) = \exists_{i,cs}\ [U(i, v, cs, u).T(i, v, cs, ns).\zeta(cs)].$$

The computation of P_ζ using the partitioned form is:

$$P_\zeta(u, v, ns) = \exists_{i,cs} [\Pi_j [u_j \equiv U_j(i, v, cs)].\Pi_k [ns_k \equiv T_k(i, v, cs)].\zeta(cs)].$$

Thus, the computation of $P_\zeta(u, v, ns)$ can be seen as the computation of the image of $\zeta(cs)$, under the transition relation $U(i, v, cs, u).T(i, v, cs, ns)$ represented in its partitioned form: $\{U_j(i, v, cs)\}, \{T_k(i, v, cs)\}$. In general, we cannot hide variables in the partitioned representation because, for this, we need to multiply out the partitions. So, we keep the variables inside the partitions and do not quantify them. However, by restricting to a specific $\zeta(cs)$ it becomes affordable to compute the product of the transitions relations and afterwards to quantify over i, therefore when we come to a specific state subset, P_ζ, we multiply the partitions, because they are restricted to the characteristic function of the subset, $\zeta(cs)$, and then we quantify over the inputs i. In other words. we say that hiding (quantification) of variables i is performed as part of the image computation.

Let $C_j(i, v, cs)$ denote the condition that is true when the j-th output of F conforms to the j-th output of S:

$$C_j(i, v, cs) = \Pi_k [O^F_{jk}(i, v, cs^F) \Rightarrow O^S_{jk}(i, cs^S)],$$

where $cs = (cs^F, cs^S)$ and the iteration is over all values k of the output O_j, e.g., for binary signals, $k \in \{0, 1\}$.

Let $C(i, v, cs) = \Pi_j \; C_j(i, v, cs)$, which is the overall condition when all outputs of F conform to the outputs of S:

$$C(i, v, cs) = \Pi_j \; \Pi_k [O^F_{jk}(i, v, cs^F) \Rightarrow O^S_{jk}(i, cs^S)].$$

Next, we compute $Q_\zeta(u, v)$, the condition on u, v when the current subset of states $\zeta(cs)$ leads to the non-conformance of outputs of F and S:

$$Q_\zeta(u, v) = \exists_{i,cs} [U(i, v, cs^F, u).\overline{C(i, v, cs)}.\zeta(cs)].$$

Since these combinations of (u, v) can lead to a non-conforming state, we exclude such transitions and map them all into a single non-accepting state *DCN* with a universal self-loop. We remove these combinations from consideration by restricting $P_\zeta(u, v, ns)$ to those (u, v) that are not contained in $Q_\zeta(u, v)$:

$$P'_\zeta(u, v, ns) = P_\zeta(u, v, ns).\overline{Q_\zeta(u, v)}.$$

Finally, the result is made complete by adding a state *DCA* (which is accepting in the final answer after complementation), with transitions from $\zeta(cs)$ into it for all (u, v) that are contained in $Q_\zeta(u, v)$.

Note that $\overline{C(i, v, cs)}$ can be represented as the sum of output non-conformance conditions for each output. Therefore, the computation of $Q_{\zeta}(u, v)$ can be done one output at a time, without computing the monolithic relation for $C(i, v, cs)$.

Regarding the representation of states and subsets of states, the following facts hold:

1. The original circuit uses logarithmic state encoding by means of state-bits.
2. During determinization, BDDs of state subsets are used as unique identifiers of states after determinization.
3. After determinization, each state of the determinized automaton is represented explicitly, without state encoding.

7.2.1.6 Validity of the Computation

The maximum prefix-closed solution (required for an FSM implementation) is computed after determinization and completion, when the set of reachable subset states, $\{\zeta(cs)\}$, are known and, for each of these, the function $Q_{\zeta}(u, v)$ has been computed, which defines the transitions from this subset state into a newly added non-accepting state *DCA*.

To perform the complementation, which is the last step in computing the maximum solution, the accepting and non-accepting states are switched. To do this, we need to determine what the accepting states are after the previous steps of computation. It is helpful to follow the computation process starting from the initial automata, F and S, and consider the two types of their states: the accepting states (labeled a) and the non-accepting states.

- S could be completed by adding a new state DC_1 which would be its only non-accepting state, because S is represented by the multi-level network and so is prefix-closed.
- In $\overline{S}$, DC_1 would be the only accepting state. (If S is non-deterministic, $\{DC_1\}$ would be the only accepting (subset) state.)
- Since F was left incomplete and is prefix-closed, all states of F are accepting.
- When forming the product $F \bullet \overline{S}$, a product state is accepting only if both states are accepting. Thus, in $F \bullet \overline{S}$, the only accepting states would be those with labels (a, DC_1), i.e., accepting in F and accepting in $\overline{S}$.
- In the subset construction, a subset state is accepting if at least one product state in the subset is accepting, i.e., of type (a, DC_1). If such occurs, then those transitions represented by $Q_{\zeta}(u, v)$ are redirected to *DCN*, since in the final answer, these states are not accepting.[5] The new completion state, *DCA*, added after the subset construction to complete it, is non-accepting for the determinized product $F \bullet \overline{S}$, but accepting in the final answer.

[5]Note that it is not necessary to compute subset states which emanate from such a state since once reached, we know that all input sequences with this prefix are not in the language of an FSM.

- Finally, in the complement of this product, which is the maximum solution, the set of accepting states are *DCA* plus all the subset states which are left in $P_\zeta(u, v, ns)$ after $\overline{Q_\zeta}$ has deleted all those that lead to *DCN*.

Thus all non-accepting subset states are mapped into *DCN*. As mentioned, this can be done because we will make the answer *prefix-closed*, a requirement for an FSM. (Thus, the X computed is the most general *prefix-closed* solution.) This requirement increases efficiency, because during the subset construction, as soon as a state of type (a, DC_1) is encountered in a subset state, $\zeta(cs)$, this state can be replaced with *DCN*. This efficiently reduces the number of subset states that emanate from such $\zeta(cs)$. Further, those states that are trimmed immediately by this need not be explored for reachability from them. This leads to a substantial trimming during the subset construction. This trimming can be performed because S is prefix-closed, and we require that X be prefix-closed also.[6]

Similarly, the existence of the single state, *DCA*, derives from the fact that F is prefix-closed. This allows for deferring the completion of F until the subset construction.

Finally, to make X into an FSM automaton, it is made input-progressive. This step is the same for both the monolithic and partitioned case and is not detailed here.

Note that in the partitioned flow, neither S nor F is made complete. In effect such completion is deferred to the all-inclusive determinization step. The validity of this is substantiated by the next section, which proves that the determinization and completion operations commute and observes that completion also commutes with complementation and product.

7.3 Completion and Determinization Commute

Theorem 7.2. *To determinize a finite automaton A, the following two procedures are equivalent:*

1. Complete(Determinize(*A*), *non-accepting*)
2. Determinize(Complete(*A*, *non-accepting*))

Proof. We show that these two procedures yield the same result by comparing subset constructions without and with pre-completion. Let DC denote the added non-accepting state. Recall that DC has a universal self-loop. It is clear that if $\{s_1, \ldots, s_k\}$ is a subset state in the automaton constructed by Procedure 1, then there must exist subset state $\{s_1, \ldots, s_k\}$ and/or $\{s_1, \ldots, s_k, DC\}$ in the automaton constructed by Procedure 2. Also, if subset state $\{s_1, \ldots, s_k\}$ or $\{s_1, \ldots, s_k, DC\}$

[6]While this trimming can be substantial, there is no avoiding that subset construction can be exponential. However, a common experience among people who have implemented subset construction, is that it can be surprisingly efficient in practice, with relatively few subset states reached, sometimes even leading to a reduction in the number of states.

exists in the resultant automaton of Procedure 2, then subset state $\{s_1, \ldots, s_k\}$ must exist in the resultant automaton of Procedure 1. Observe that

- DC state is non-accepting. A subset state is accepting if and only if one of its states is accepting. Thus, state $\{s_1, \ldots, s_k\}$ is accepting if and only if state $\{s_1, \ldots, s_k, DC\}$ is accepting.
- Under any transition condition, the successor states of $\{s_1, \ldots, s_k\}$ are the same as those of $\{s_1, \ldots, s_k, DC\}$ except that the latter contains the DC state.

Applying induction on the state space with the above two facts, it follows that subset states $s_1, \ldots, s_k$ and $s_1, \ldots, s_k, DC$ are equivalent. On the other hand, if a subset state $s_1, \ldots, s_k$ or $s_1, \ldots, s_k, DC$ can transition to the singleton subset state $\{DC\}$ under some transition condition, then all of the member states $s_1, \ldots, s_k$ must be incomplete under this transition condition. Therefore, it is immaterial whether the completion is done before or after the determinization. □

Other trivial propositions state that completion commutes with both the complementation and product operations. Corollary 7.3 follows from these observations and Theorem 7.2.

Corollary 7.3. *Let X be the solution obtained when F and S are not completed and let Y be the solution when F and S are made complete first. Then the languages of X and Y are identical.*

7.4 Experimental Results

In this section, an implementation using the partitioned representations is compared with the implementation using the monolithic representations. In effect, the partitioned implementation performs only the determinization procedure, which subsumes all other steps, as described above. In the case of the monolithic representations, the completion of S is done first, then the intermediate product is derived, followed by hiding and determinization, performed in a traditional way.

The examples for this experiment were derived from FSM benchmarks by the operation called *latch splitting*. It is explained in Chaps. 11 and 13. It is a syntactic transformation of a sequential circuit into two circuits, one containing a subset of the latches of the original circuit and the other containing the rest. One of these becomes the fixed component, F, in the language solving problem while the other represents a particular solution, X_P, for the "unknown" component. The sequential behavior of the original circuit is used as the specification. We compute the largest FSM solution called CSF (complete sequential flexibility). Since X_P, is a particular solution, it is contained in the CSF.

After the computation of CSF, it was formally verified by checking:

1. $X_P \subseteq CSF$, and
2. $F \bullet CSF \subseteq S = F \bullet X_P$.

Table 7.1 Comparison of partitioned and monolithic representations

Name	$i/o/cs$	F_{cs}/X_{cs}	States(X)	Part,s	Mono,s	Ratio
s510	19/7/6	3/3	54	0.3	0.2	0.7
s208	10/1/8	4/4	497	0.4	0.8	2.0
s298	3/6/14	7/7	553	0.9	2.7	3.0
s349	9/11/15	5/10	2,626	37.7	810.3	21.5
s444	3/6/21	5/16	17,730	25.9	CNC	–
s526	3/6/21	5/16	141,829	276.7	CNC	–

Table 7.1 lists the results for several examples. The columns show the example name (column "Name"), the number of inputs, outputs and latches (column "$i/o/cs$"), the number of latches in the fixed part and in the current implementation of the unknown (column "F_{cs}/X_{cs}"), the number of states in the CSF (column "States(X)"), the runtime in seconds of the partitioned algorithm (column "Part") and runtime in seconds using monolithic representations (column "Mono"), and finally the ratio of these runtimes (column "Ratio").

The algorithms were implemented in the BALM. The measurements were made on a Windows XP computer with a 1.6 GHz CPU.

The results show that the partitioned method is more efficient (except for very small examples) with efficiency increasing as the problem size increases. At some point, the monolithic method can not complete (CNC) when the intermediate automata become relatively large. Of course, even the partition method will run out of steam at a certain point, but note that on one example it could handle over 140,000 states in the CSF.

7.5 Conclusions

In this chapter, a strategy was presented for solving language equations when the fixed component and the specification are represented using multi-level deterministic sequential networks. In such cases, a partitioned representation derived from the original network can be used to perform all the operations needed to compute the maximum solution. The operations required can be re-formulated in terms of image computations, which can take advantage of the advances made in the efficiency of these computations over the last decade. The new methods allow for faster solution of larger problem instances. We emphasize that the techniques described, rely heavily on the fact that both automata, F and S, are represented initially by multi-level sequential networks and hence are prefix-closed, and that the desired answer is also prefix-closed.

It was proved also, that not completing F in the algorithm leads to the same answer. This has two important consequences. First, it is not necessary to form the completion conditions before computing the product or its determinization. This saves a lot of computing. Second, completion of F can lead to a non-deterministic product after hiding, while leaving F incomplete leads to a deterministic product after hiding. Thus a subset construction is avoided when F is not completed.

Of course, computing the CSF is only one step in sequential synthesis. Finding an optimum sub-solution of the CSF remains an outstanding problem for future research.

Problems

7.1. Representing Automata by MV Networks
Manipulation of finite automata is performed using the graph representation, as discussed in Chap. 6.

However automata can also be represented and manipulated using multi-valued multi-level networks that implement the next-state logic F, the latches (one or more multi-valued latches according to the encoding) and the acceptance unction Acc. F is the next-state logic that depends on the inputs and present-state variables and produces the next-state variables. If the automaton is non-deterministic F too is non-deterministic. Acc is a binary node producing value 1 if and only if the present-state is accepting; it depends only on the present-state. In practice we restrict ourselves to representing deterministic automata, because non-deterministic ones introduce complications[7] in the network representation.

Describe how to represent automata by MV networks. Discuss the various options for encoding and logic optimization algorithms.

Note that the unused codes should be stored in a specific network, call it EXDC (external don't cares) network, and exploited as such when appropriate. The EXDC network must be updated when the latches change.

7.2. Manipulating Automata by MV Networks
In Problem 7.1 we represented a deterministic automaton by a MV network with multiple latches.

Once an automaton is encoded as a network, the next task is to express the following operations on automata as operations on MV networks, i.e., to provide algorithms that perform these operations by working directly on the data structures representing the MV networks:

(a) Operation *complement* of an automaton.
(b) Operation *product* of two automata.
(c) Operation *variable projection* of an automaton.
(d) Operation *determinization* of an automaton.

---

[7] Say that an MV network has two binary latches and that under some input both latches produce both 1 and 0. Does this situation correspond to the subset of states $(00, 11)$, or $(01, 10)$ or $(00, 11, 10)$? To avoid loss of information we should introduce additional input variables to distinguish the situations. This is an unexplored area of synthesis.

Hint. Operations projection and determinization should be merged together into a single operation, since projection may introduce non-determinism and we are restricting ourselves to the representation of deterministic automata. This is a difficult step and there is no established good procedure for doing it.

(e) Realize the operation *prefix-closure* for an automaton on its corresponding MV network.

Hint. We are supposed to remove non-accepting states (and the transitions from them). The problem is that we cannot remove non-accepting states and still keep the MV network well-defined (specified under any input). Add the non-accepting states to the EXDC network (this is consistent with the fact that their next states can be any states).

(f) Realize the operation *progressiveness* for an automaton on its corresponding MV network.

Hint. We are supposed to remove the accepting states that have no next state under some input pattern. This amounts to check all the states represented by the *Acc* function: if for a given present state there is an input pattern under which its next state is in the EXDC network, then this present state should be removed from *Acc* and added to the EXDC network; iterate until nothing changes.

(g) Realize the operation *state minimization* for an automaton on its corresponding MV network.

Chapter 8
The Software Package BALM

8.1 Introduction

Finite automata and their languages are well-studied topics since the early development of computation theory. Traditional implementations of automata manipulations are based on explicit state representation, and are limited to automata with a few thousand states. The manipulation of automata became more practical with the advent of efficient symbolic techniques based on binary decision diagrams (BDDs), satisfiability (SAT) solvers and AND-INVERTER graphs (AIGs).

Based on these techniques, the BALM (Berkeley Automata and Language Manipulation) package provides an experimental environment for more efficient manipulation of finite automata in various application domains, e.g., synthesis, verification, control, etc. The BALM package includes the most typical automata operations, such as determinization and state minimization, as well as a few visualization capabilities, which rely on the powerful graph visualization software Graphviz [49]. The applicability of BALM to finite-state machine synthesis is illustrated by its use in solving several problems that are formulated as unknown component problems using language equations. BALM is designed like SIS and MVSIS to have a command line on which a command can be entered. A sequence of commands can be read from a file, called a script. In the appendix, we list and briefly explain the presently implemented commands of BALM.

Compared with other automata manipulation utilities, such as Grail [117], FSA [43], FIRE [41] and many others (see also [144]), BALM provides an environment particularly suited for language equation solving. It can be used in component-based sequential hardware synthesis. It is useful as an educational tool, which can further the understanding of finite automata. In addition, it may be useful in sequential system optimization, because it can interface with other programs based on the logic file formats, such as `BLIF-MV` or `BLIF`, which are commonly used to represent designs in academic logic synthesis tools. With BALM, it is possible to extract sequential flexibilities in implementing a hardware finite state machine. We will illustrate some of these applications.

T. Villa et al., *The Unknown Component Problem: Theory and Applications*,
DOI 10.1007/978-0-387-68759-9_8,

In addition to most of the traditional explicit graph enumeration algorithms for automata manipulations, BALM provides computation algorithms based on state-of-the-art symbolic techniques. This allows most manipulations of automata to be done implicitly or using a mixture of explicit and implicit techniques. With these, the system can handle millions of states. For example, complementation (determinization) of nondeterministic automata, which is an operation based on hybrid representations, can produce deterministic automata with 100 K states in several seconds on modern computers.

8.2 Describing FSMs and Automata

BALM supports two file formats, the `BLIF-MV` (and `BLIF`) format for describing FSMs as multi-level netlists of logic, and the `AUT` format for describing automata. The `AUT` format is a restricted form of `BLIF-MV`.

8.2.1 The BLIF-MV Format

`BLIF-MV` was created for VIS [15, 16], a verification package, and is an extension of `BLIF` (Berkeley Logic Interchange Format) to efficiently specify sequential systems in the form of multi-level multi-valued (possibly nondeterministic) logic networks. Compared to `BLIF`, `BLIF-MV` allows (1) multi-valued signals, and (2) nondeterministic nodes[1], which can be used to model nondeterministic systems. For instance, this is useful in the early stages of design development, when a sequential system may contain non-determinism because some of its components are still to be specified, and only an abstraction is known. `BLIF-MV` supports multi-valued variables, which often simplifies the specification of a system. Each multi-valued variable is given a size and optionally a list of value names. Currently, the number of values of the multi-valued variables in the implementation of BALM is limited to 32; however, the number of values of multi-valued variables denoting states variables in the `AUT` format is not limited.

The semantics of `BLIF-MV` is defined using a combinational/sequential concurrency model. There are three basic primitives: variables, tables (nondeterministic nodes), and latches. A variable can take values from a finite domain. A relation defined over a set of variables is represented using a table. A table has only one output and any number of inputs. A particular variable can be designated as the output of at most one table. Several tables are conceptually inter-connected if they share the same variable name, i.e. if there is a common name at the output of one

[1] Such a node generates, for a given input, a randomly selected output value from a specified subset of output values.

table and the inputs of the other tables. Thus a named variable is conceptually a wire. If a table is deterministic and binary, it can also be thought of as a logic gate. A latch is the state-holding element of BLIF-MV. It has only one input. Its output has the same set of allowed values (domain) as its input. A set of initial values is associated with every latch; they must be a subset of the domain of the latch. A state is an assignment of values to the latches of a design, where a value assigned must be in the domain of the latch. An initial state is a state where every latch takes one value from its set of initial values. A latch can have more than one initial state in general.

The semantics are as follows. At every time point, the system is in some state (each latch has a value). At every clock tick, each latch updates its output value to be its current input value. These outputs then propagate through tables in a topological order until all the wires have a consistent set of values, i.e. each output value of a table is one of the set of allowed values for the set of input values on the inputs to the table. If a latch is encountered during the propagation, i.e., an output of a table is an input to a latch, the propagation process is stopped. Note that because of nondeterminism, given a single state in the latches, there may be several consistent sets of values on the wires. These semantics can be seen as a simple extension of the standard semantics of synchronous single-clocked digital circuits. In fact, if every table is deterministic and every latch has a single initial state, the two semantics are exactly equal. Thus, the only differences are in the interpretation of nondeterministic tables and latches with multiple initial states.

In BALM, the command `read_blif_mv` reads a BLIF-MV (or BLIF) description, and then sets up a corresponding internal data structure to represent the multi-valued nondeterministic network. The `write_blif_mv` command writes a BLIF-MV description to a file. Although the BLIF-MV format was designed as a logic exchange format, simple examples in BLIF-MV are readable and can exhibit some degree of clarity.

Example 8.1. We show an example of the BLIF-MV description of a multi-valued network with one internal node and one latch, where all signals are 3-valued:

```
.model blif_mv_example
.inputs a
.outputs f g
.mv a,f,g 3
.table a g -> f
.default 0
1 0 1
0 2 2
.latch f g
.reset g
2
.end
```

For a more detailed treatment of the BLIF-MV format and some examples, see [75].

8.2.2 The AUT Format

The AUT format was designed to describe finite automata in a simple two-level form. It is essentially a restricted subset of BLIF-MV. The adopted restrictions allowed for a simplified version of the BLIF-MV parser; the restrictions are not an inherent part of any theory, and could be relaxed if future applications dictate. The unrestricted BLIF-MV is also used in BALM to represent the FSMs in the form of multi-valued multi-level nondeterministic networks, as described in the previous section.

The restrictions of AUT are described below. The example that follows illustrates many of the points:

1. Only non-hierarchical BLIF-MV specifications are allowed.
2. An AUT file must have exactly one latch and exactly two combinational tables, one describing the next-state relation and one describing a single output named "Acc". Multi-level netlist descriptions of the next-state relation are not supported.
3. The latch's output variable is named "CS" and its input variable is named "NS". They have the same domain whose size must be equal to the number of states of the automaton. The reset value of the latch is the initial state of the automaton. Currently, only automata with one initial state can be used. This data is specified using the .mv directive.
4. The next-state table (output is NS) must have the automaton inputs listed on the ".table" line, followed by CS. The last variable on the line must be NS, the output of the table.
5. In each line of the next-state table (which specifies a multi-valued cube), variables CS and NS can have only one specific state value; for example, "s33"or "stateABC". Multi-valued state literals of the type "(s33, s35, s37)" are not allowed in the next-state table. The don't-care literal "-" cannot be used for the CS and NS variables. All the values of CS must be used in the table at least once, including the ".default" line. There are no such restrictions on how the multi-valued inputs of the automaton are specified in the table. Any literals of these variables, including the don't-care literal, can be used in the next-state table.
6. The BLIF-MV specification must have exactly one binary primary output. Its name is must to be "Acc" and is defined using the other combinational table of the AUT file. It has only one input, CS. The purpose of this node is to specify which of the automaton states are accepting. Thus Acc = 1 exactly when CS has a value equal to one of the accepting states.
7. The default line of the Acc table can be in one of the following forms:

Case 1: all states are accepting

```
.table -> Acc
1
```

This specifies that Acc is the constant 1.

Case 2: only one state is accepting

```
.table CS -> Acc
.default 0
<acc_state> 1
```

This specifies that state "acc_state" is the only accepting state.

Case 3: only one state is non-accepting

```
.table CS -> Acc
.default 1
<non_acc_state> 0
```

This specifies that state "non_acc_state" is the only non-accepting state

Case 4: several accepting states

```
.table CS -> Acc
.default 0
```

(comma-separated list of accepting states in parentheses) `1`

or

```
.table CS -> Acc
.default 1
```

(comma-separated list of non-accepting states in parentheses) `0`

Example 8.2. We show an example of a simple 3-state automaton with accepting state "DC" represented using the AUT format:

```
.model spec
.inputs i o
.outputs Acc

.mv CS, NS  3  a b DC #there are 3 states named a b and DC
.mv i 3 #input i has 3 values. By default their names are 0,1,2

.latch NS CS #declares one latch with input NS and output CS
.reset CS #its initial value is a
a

.table CS ->Acc #the only accepting state is DC
.default 0
DC 1

.table i o CS ->NS #the next-state table
.default DC
(1,2)  1   a   a #if i = 1 or 2, and o =1 and CS=a, then NS=a
-      0   a   b
0      1   b   a

.end
```

The `.default` keyword in the `.table` construct is a shortcut for the following cases:

```
0      1   a   DC
(1,2)  1   b   DC
-      0   b   DC
-      -   DC  DC
```

In BALM, the command `read_blif_mv` can also read an `AUT` file because `AUT` is a subset of `BLIF-MV`. In such a case, the file is interpreted as a multi-valued network and not as an automaton (the number of states in this case should not exceed 32). There is no separate command to read in an automaton as an automaton, since the automata manipulation commands always read and write to `AUT` files; the input automata file name(s) are on the command line followed by the file name where the result will be written. Thus, there is no need for BALM to have separate commands to read and write automata, say `read_aut` and `write_aut`, because there is no notion of the current automaton, and instead each command that operates on automata reads and writes the automaton of interest in a specific file with extension `.aut`.

8.2.3 Comparing the BLIF-MV and AUT Formats

The difference between `BLIF-MV` and `AUT` (which is a modified `BLIF-MV`) is that `BLIF-MV` (as interpreted by MVSIS/BALM) can only have multi-valued variables up to 32 values. Meanwhile, `AUT` can have an arbitrary number of values. BALM uses a dedicated `AUT`-parser to bypass the limitations of the `BLIF-MV`-parser in MVSIS and be able to parse multi-valued variables with more than 32 values. Therefore, using `AUT` we can represent FSMs and automata with any number of states.

The multi-valued network flow in MVSIS and the automata/FSM flow in BALM are organized differently. MVSIS is based on the notion of the current network. So there is a command (`read_blif_mv`) which just reads the network in from file, making it the current network. BALM inherits the notion of current multi-valued network from MVSIS, but does not rely on the notion of the current automaton/FSM, because each command is made to read its own input and produce its own output. This is why there is no special command to read the automaton as automaton. Of course, we can read the automaton as a multi-valued network (provided it has no more than 32 states), but then we are working with MVSIS, not BALM.

Example 8.3. Suppose that we want to describe a counter FSM with 32 states whose description in the `kiss` format [125] is the following:

```
.i 1
.o 1
.s 32
```

```
.p 64
0 st0 st0 0
1 st0 st1 1
0 st1 st1 0
1 st1 st2 1
...
0 st30 st30 0
1 st30 st31 1
0 st31 st31 0
1 st31 st0 1
.end
```

We may describe it in the BLIF-MV format (say in the file counter32.mv)[2]:

```
.model counter32.mv
.inputs i
.outputs SP o
.mv SF, SP 32 st0 st1 ... st31
.table i SP ->SF
0 st0 st0
1 st0 st1
0 st1 st1
1 st1 st2
...
0 st31 st31
1 st31 st0
.latch SF SP
.reset SP
0
.table i SP ->o
0 st0  0
1 st0  1
0 st1  0
1 st1  1
...
0 st31 0
1 st31 1
.end
```

Otherwise, we could describe it in the AUT format as follows (say in the file counter32.aut):

```
.model counter32.aut
.inputs i o
.outputs Acc
.mv CS, NS 32 st0 st1 ... st31
.latch NS CS
.reset CS
0
```

---

[2]Lines introducing mv variables like .mv SF, SP 32 st0 st1 ... st31 stand for the full list of symbolic names of the values of the mv variables, e.g., .mv SF, SP 32 st0 st1 st2 st3 st4 st5 st6 st7 st8 st9 st10 st11 st12 st13 st14 st15 st16 st17 st18 st19 st20 st21 st22 st23 st24 st25 st26 st27 st28 st29 st39 st31.

```
.table Acc #all states are accepting
1
.table i o CS ->NS
0 0 st0 st0
1 1 st0 st1
0 0 st1 st1
1 1 st1 st2
...
0 0 st31 st31
1 1 st31 st0
.end
```

The latter format would apply also if the counter FSM would have say 33 states, whereas the former format would not. Consider the following counter with 33 states:

```
.i 1
.o 1
.s 33
.p 66
0 st0 st0 0
1 st0 st1 1
0 st1 st1 0
1 st1 st2 1
...
0 st31 st31 0
1 st31 st32 1
0 st32 st32 0
1 st32 st0 1
.end
```

The previous AUT `description` would be extended easily as follows (say in the file `counter33.aut`):

```
.model counter33.aut
.inputs i o
.outputs Acc
.mv CS, NS 33 st0 st1 ... st31 st32
.latch NS CS
.reset CS
0
.table Acc #all states are accepting
1
.table i o CS ->NS
0 0 st0 st0
1 1 st0 st1
0 0 st1 st1
1 1 st1 st2
...
0 0 st31 st31
1 1 st31 st32
0 0 st32 st32
1 1 st32 st0
.end
```

```
.model spec
..inputs
..outputs out
..mv out 3 OK notOK done
..table ->out
.OK
.done
..end
```

Fig. 8.1 BLIF-MV description of the automaton discussed in Example 8.4

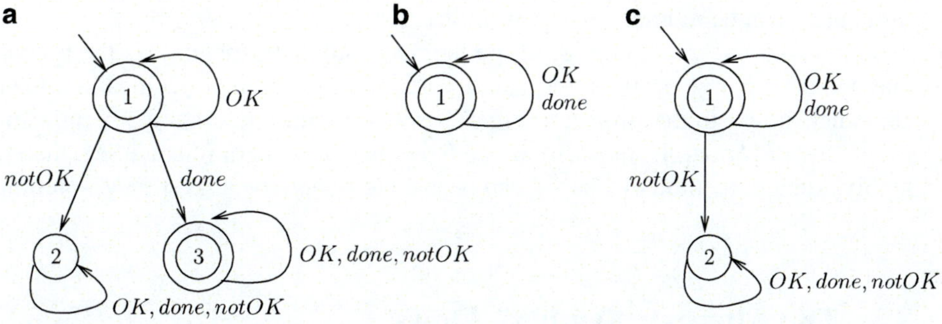

Fig. 8.2 Illustration of Example 8.4. (**a**) Automaton Structure 1; (**b**) Automaton Structure corresponding to BLIF-MV description in Fig. 8.1; (**c**) Automaton Structure 2

8.2.4 Describing an Automaton Using BLIF-MV

The BLIF-MV format does not have the notion of accepting states, and therefore has a limited ability to describe an automaton. By default it describes an FSM and hence all states are accepting. The only possibility for having a non-accepting state is adding a state to an incomplete automaton when it is completed. Thus BLIF-MV can only be used to describe automata where all states are accepting except one. The reason why this is still useful is that if the specification automaton can be described in BLIF-MV, then the efficient method of Chap. 7 can be used to solve for an FSM solution. In many examples, having only one non-accepting state is enough, as shown in Example 8.4.

Example 8.4. Suppose that we want to specify in BLIF-MV the automaton with 3 states shown in Fig. 8.2a.

The states 1 and 3 are accepting states and 2 is a non-accepting state. Consider the BLIF-MV description shown in Fig. 8.1.

The table for out does not produce the output value notOK. Consider what happens when this FSM is converted into an automaton and out becomes the input. Since it has no latches, it is an FSM with only one state, as shown in

Fig. 8.2b. The transitions from this state to a next state are defined only for `out =OK` and `out=done`, but not for `out = notOK`. As an automaton, it is incompletely defined. When the automaton is completed it will have two states as shown in Fig. 8.2c, where state 2 is non-accepting.

Note that this is not exactly the same as the three-state automaton in Fig. 8.2a, because the two accepting states have been merged into one state. However, for some applications, this can be done. For example, suppose that, when the fixed part produces `out = done` or `out = notOK`, then no further changes in state happen. Then it is not important what happens after states 2 or 3 in Fig. 8.2 are entered. Thus, any transition added to state 3 in Structure 1 will be irrelevant. If the transition $3 \rightarrow 2$ is added, then state 3 becomes equivalent to state 1 and hence Structure 1 can be minimized to that of Structure 2.

In summary, if the fixed part of the language solving problem has the property that no state change occurs after a sink state is entered in the specification, then a construct similar to that shown in Fig. 8.1 can be used to describe the specification in `BLIF-MV` format. If, in addition, the fixed part is a deterministic FSM, then the highly efficient method of Chap. 7 can be used to obtain the largest FSM solution.

8.3 Two Alternate Synthesis Flows

BALM supports two different flows for solving language equations or manipulating languages. The languages are represented in either `BLIF-MV` or `AUT` formats[3]. In the former case it means that we start from a sequential circuit from which an STG can be extracted, whereas in the latter case we start already from a given STG.

1. The first flow deals with FSMs where typically one has a fixed part F and a specification S given as FSMs. This flow is oriented towards solving for the unknown component X the equation $F \bullet X \subseteq S$. The same sequence of operations could be done with the second type of flow using only `AUT` files, but this FSM flow takes advantage of the special features of F, S, and X to provide a very efficient implementation, as discussed in detail in Chap. 7. The entire flow is embodied in a single command, `solve_fsm_equ`, which produces the largest FSM solution X in `AUT` format. This command takes advantage of the fact that F and S are deterministic FSM automata and X is required to be an FSM automaton[4].
2. The second flow deals directly with automata where each step consists of reading in input `AUT` files, manipulating them according to the command, and writing

[3]Notice that technically the `AUT` format is just a special case of the `BLIF-MV` format.

[4]Precisely, the initial sequential circuits F and S yield deterministic FSMs whose automata define the language equation for which we look for a solution that is an FSM automaton, i..e., a prefix-closed input-progressive automaton.

out a resulting automaton in AUT format. In addition, there is a command which extracts an equivalent automaton from a BLIF-MV file: the command is extract_aut. A typical script executing equation solving is the following:

```
read_blif_mv fixed.mv
extract_aut fixed.aut
complement spec.aut specc.aut
product fixed.aut specc.aut product.aut
support u product.aut product_supp.aut
complement product_supp.aut x.aut
progressive -i n x.aut xfsm.aut
minimize xfsm.aut xfsm_min.aut
```

Note that each command, except for read_blif_mv, reads in an automaton file and writes out an automaton file. Here u is a list of support variables and n is the number of inputs that the solution will have as an FSM.

If an automaton would be already available as fixed.aut (say written directly in the AUT format) then there would be no need to run first the read_blif_mv and extract_aut commands and the previous flow would reduce to:

```
complement spec.aut specc.aut
product fixed.aut specc.aut product.aut
support u product.aut product_supp.aut
complement product_supp.aut x.aut
progressive -i n x.aut xfsm.aut
minimize xfsm.aut xfsm_min.aut
```

8.3.1 Some Specialized Automata Operations

In BALM, command print_nd_states checks if an automaton is deterministic. A nondeterministic automaton can be determinized using the command determinize. This uses a subset construction.

Determinizing a nondeterministic automaton is a step usually performed before complementation since complementing a deterministic automaton A can be easily achieved by inverting the acceptance condition of states. In BALM, an automaton A can be complemented using command complement where a determinization step is automatically performed first if A is nondeterministic.

A state is incomplete if there exists some input assignment, under which the next state transition is undefined (i.e., there is no next state under that input assignment). An automaton is said to be incomplete if it has at least one incomplete state. An incomplete automaton can be completed by adding a single non-accepting don't-care state with a self-loop transition under any input assignment (thus the don't-care state is a sink state). All missing input assignments of an incomplete state are assigned as transitions from that state to this don't-care state. Note that the automata before and after completion both accept the same language. In BALM, an incomplete automaton can be completed using command `complete`. By default, this command adds a non-accepting don't-care state. In some applications, it is necessary to add an accepting don't-care state, which is done using command `complete -a`.

An automaton obtained from an FSM (by combining inputs and outputs) is prefix-closed and input-progressive. Given an automaton A with the specification of input and output variables, BALM is capable of trimming A to be the automaton of an FSM.

An automaton A is prefix-closed if any prefix of an accepting string in $L(A)$ is also in $L(A)$. A deterministic automaton can be trimmed to be prefix-closed by the following three steps: (a) removing all non-accepting states and transitions into them and from them, (b) removing states unreachable from the initial state, (c) completing the resulting automaton. BALM provides command `prefix` to trim an automaton to be prefix-closed.

A state of an automaton is progressive with respect to a set of variables U, called U-progressive, if under any valuation of u at least one of its next states is accepting. An automaton is progressive with respect to U, if all of its states are U-progressive. An automaton can be trimmed to be progressive with respect to U by iteratively deleting states that are not U-progressive. BALM can trim an automaton to be progressive by command `progressive`. The number of variables that are to be considered as inputs to U needs to be specified on the input line, e.g., `progressive -i 5` specifies that the first 5 input variables of the automaton are considered as inputs, the remaining as outputs. By convention, inputs must occur first in the list of variables. Otherwise, command `support` can be used to reorder the list.

Thus the command `prefix; progressive` can be used to convert a general automaton into an FSM automaton (if one exists). In fact, `progressive` is defined so that if the input automaton is not prefix-closed, it will be made prefix-closed automatically. These operations will trim some states and transitions, so that the result may not be a well-defined FSM, i.e., for every input minterm there exists a next state. This can happen during language solving where we first derive the most general solution automaton, from which we want to extract an FSM automaton. If the trimming process results in a not well-defined FSM, then this implies that there is no FSM solution. Also, BALM can further constrain the synthesized FSM to be of Moore type. This may result in more trimming and also may lead to a not well-defined machine.

In composing two FSMs, it is sometimes necessary to rearrange (rename, reorder, create, or hide variables) input and output signals. BALM supports these rearrangements by one command `support`.

In BALM, a deterministic finite automaton can be state-minimized using command `minimize` based on the Myhill-Nerode theorem [63]. In this, the full behavior is preserved, just the number of states is reduced. In contrast, BALM provides a heuristic algorithm (command `dcmin`), which can be applied to a nondeterministic finite automaton. In this, some behavior may be eliminated (the resulting language is a subset of the original language). This is used when the largest solution of a language equation is computed first, which contains all possible solutions, whereas we only want one small solution. A discussion of the use of `dcmin` can be found in Sect. 14.3.

Many verification problems in state-transition systems can be reduced to checking language containment, which tests if the language of one automaton is contained in that of another. The checking can be accomplished by the product and complement operations. In BALM, language containment checking is performed using command `contain`, which will report if the two automata specified on the command line are related by language containment (or are equivalent). If the automata are not equivalent, a counter-example can be produced.

8.4 Appendix: Commands in BALM

BALM can be downloaded from the following site:
http://embedded.eecs.berkeley.edu/mvsis/balm.html.

The following list contains a one line summary of all the commands available in BALM.

Automata manipulation commands:

- `complement`: complement an automaton (a nondeterministic automaton will be automatically determinized first)
- `complete`: complete an automaton by adding a don't-care state
- `contain`: check language containment of two automata (checking is automatically aborted is at least one automaton is nondeterministic)
- `dcmin`: minimize the number of states by collapsing states whose transitions into care states are compatible
- `determinize`: determinize an automaton
- `minimize`: minimize the number of states of a deterministic automaton
- `moore`: trim an automaton to contain Moore states only
- `prefix`: leave only accepting states that are reachable from initial states
- `product`: build the product of two automata
- `progressive`: leave only accepting and complete states that are reachable from initial states
- `support`: change the input variables of an automaton[5]

[5] Two caveats about using `support`:

1. One must declare explicitly the number of values of input variables with more than two values.
2. It cannot handle an automaton with only one state; a work-around is to define an automaton with two equivalent states.

Automata viewing commands:

- `plot_aut`: visualize an automaton using DOT and GSVIEW
- `print_lang_size`: compute the number of I/O strings accepted (within the string length set by the argument `-l`) by the maximum prefix-closed sub-automaton of an automaton
- `print_nd_states`: print information about nondeterministic states of an automaton
- `print_stats_aut`: print statistics about an automaton
- `print_support`: print the list of support variables of an automaton

I/O commands:

- `read_blif`: read the current network from the `BLIF` file
- `read_blif_mv`: read the current network from the `BLIF-MV` file
- `write_blif`: write the current network in the `BLIF` format
- `write_blif_mv`: write the current network in the `BLIF-MV` format

Miscellaneous commands:

- `alias`: provide an alias for a command
- `echo`: echo the arguments
- `help`: print the list of available commands by group
- `history`: a UNIX-like history mechanism inside the BALM shell
- `ls`: print the file names in the current directory
- `quit`: exit BALM
- `source`: execute commands from a file
- `time`: provide a simple elapsed time value
- `unalias`: removes the definition of an alias

MV network commands:

- `extract_aut`: extract the state-transition graph from the current network as an automaton
- `latch_expose`: make latch outputs visible as POs of the current network
- `latch_split`: split the current network into two networks by dividing latches and the related combinational logic; generates synthesis and verification scripts assuming that one part is fixed and another part is unknown
- `solve_fsm_equ`: solve language equation $F \bullet X \subseteq S$ using the method discussed in Chap. 7. F and S must be given in `BLIF-MV` format and must be deterministic.

Network viewing commands:

- `print`: print multi-valued sum-of-products representation of nodes
- `print_factor`: print algebraic factored form of nodes
- `print_io`: print fanins/fanouts of nodes
- `print_latch`: print the list of latches of the current network

- `print_level`: print nodes in the current network by level
- `print_nd`: print the list of nondeterministic nodes in the current network
- `print_range`: print the numbers of values of nodes
- `print_stats`: print network statistics and report the percentage of nodes having each representation

Problems

8.1. Consider the context automaton *ma* defined by the following BLIF-MV description:

```
.model ma
.inputs i v u o
.outputs Acc

.mv CS, NS 2 a b
.mv i 2 i1 i2
.mv v 2 v1 v2
.mv u 2 u1 u2
.mv o 3 o1 o2 o3

.latch NS CS
.reset CS
a

.table CS -> Acc
.default 1

.table i v u o CS -> NS
i1 v1 u1 o1 a a
i2 v2 u1 o2 a a
i1 v2 u1 o2 a a

i1 v1 u1 o3 a b
i1 v1 u2 o1 a b

i1 v1 u2 o3 b b

i2 v1 u1 o1 b a
i1 v2 u1 o2 b a

.end
```

and the specification automaton mc defined by the following `BLIF-MV` description:

```
.model mc
.inputs i o
.outputs Acc

.mv CS, NS 2 s1 s2
.mv i 2 i1 i2
.mv o 3 o1 o2 o3

.latch NS CS
.reset CS
s1

.table CS -> Acc
.default 1

.table i o CS -> NS
i1 o2 s1 s1
i1 o3 s1 s1
i2 o2 s1 s1

i1 o1 s1 s2

i1 o3 s2 s2
i2 o1 s2 s2

i1 o2 s2 s1
i2 o2 s2 s1

.end
```

Use BALM to solve the equation $ma \bullet mx \subseteq mc$, where mx has support v, u. Compute the following operations, invoking the appropriate commands in BALM, and at each step plot the resulting automaton:

1. Compute mc_comp, the complement of mc.
2. Compute $ma_*(mc\_comp)\_prod$, the product of ma and mc_comp.
3. Compute $ma_*(mc\_comp)\_prod\_supp$, the projection to the support v, u of the previous automaton $ma_*(mc\_comp)\_prod$.
4. Compute mx, the complement of the previous automaton $ma_*(mc\_comp)\_prod\_supp$, i.e., the largest automaton solution of the equation ($mx = \overline{(ma \bullet \overline{mc})}$).

8.2. Perform using the appropriate commands of BALM the following operations, and at each step plot the resulting automaton:

1. Compute mx_prog, the largest progressive automaton contained in mx.
2. Compute mx_prog_min, the automaton obtained by the (state) minimization of mx_prog.

3. Compute *mx_min*, the automaton obtained by the (state) minimization of mx.
4. Compute *mx_min_prog*, the largest progressive automaton contained in *mx_min*.
   Compare *mx_prog_min* vs *mx_min_prog*, and comment on their differences if any.
5. Compute *mx_min_prog_min*, the automaton obtained by the (state) minimization of mx.
   Compare *mx_prog_min* vs. *mx_min_prog* vs. *mx_min_prog_min*, and comment on their differences if any.

8.3. Consider the context FSM M_A and the specification FSM M_C whose descriptions in `kiss` format are, respectively, the following:

```
.i 2
.o 2
.s 2
.p 8
.r a

i1 v1 a a u1 o1
i1 v1 a b u1 o3
i1 v1 a b u2 o1
i1 v2 a a u1 o2
i2 v2 a a u1 o2

i1 v1 b b u2 o3
i1 v2 b a u1 o2
i2 v1 b a u1 o1
.end
```

and

```
.i 1
.o 1
.s 2
.p 8
.r s1

i1 s1 s1 o2
i1 s1 s1 o3
i2 s1 s1 o2
i1 s1 s2 o1

i1 s2 s2 o3
i2 s2 s2 o1
i1 s2 s1 o2
i2 s2 s1 o2
.end
```

Using the appropriate commands in BALM, find the largest FSM solution of the FSM equation $M_A \bullet M_X \subseteq M_C$, where M_X has inputs $U = \{u1, u2\}$ and outputs $V = \{v1, v2\}$. Analyze carefully the result of each step.

8.4. Given the context FSM M_A and the specification FSM M_C consider the underlying automaton A and C whose descriptions in `aut` format are, respectively, the following:

```
.model es43ma
.inputs i u o
.outputs Acc

.mv i 2 i1 i2
.mv u 2 u1 u2
.mv o 2 o1 o2
.mv CS, NS 3 a b c

.latch NS CS
.reset CS
a

.table ->Acc
1

.table i u o  CS ->NS
i2 u1 - a a
i1 u1 - a b
i2 u1 - b a
i1 u2 - b c
-  u1 - c a
.end
```

and

```
.model es3mc
.inputs i o
.outputs Acc

.mv CS, NS 6 1 2 3 4 5 6

.mv i 2 i1 i2
.mv o 2 o1 o2

.latch NS CS
.reset CS
1
```

```
.table CS -> Acc
.default 1

.table i o CS -> NS
i1 o1 1 4
i2 o2 4 1
i1 o2 4 5
-  o2 5 1
i2 o1 1 2
i2 o2 2 1
i1 o2 2 3
i2 o1 3 2
i1 o2 3 6
-  o1 6 2
.end
```

Using the appropriate commands in BALM, find the largest automaton solution of the automata equation $A \bullet X \subseteq C$, where X has inputs u, o.

One can relate easily these results to the solution of $M_A \bullet M_X \subseteq M_C$, where M_A has inputs i and outputs u, M_C has inputs i and outputs o, M_X has inputs u and outputs o.

Notice that in this example every reduction of the largest solution is a solution.

Chapter 9
Fundamental Operations in BALM

In this chapter, we illustrate some of the basic automata operations using small examples so that the results can be plotted with reasonable readability. All of the examples were computed using BALM commands and the BALM plot command, `plot_aut`, to display the automata graphically. We illustrate most of the operations below using common examples derived from the file `S27.blif`, shown below

```
.model s27.bench
.inputs G0 G1 G2 G3
.outputs G17
.latch     G10 G5  0
.latch     G11 G6  0
.latch     G13 G7  0
.names G11 G17
0 1
.names G14 G11 G10
00 1
.names G5 G9 G11
00 1
.names G2 G12 G13
00 1
.names G0 G14
0 1
.names G14 G6 G8
11 1
.names G1 G7 G12
00 1
.names G12 G8 G15
1- 1
-1 1
.names G3 G8 G16
1- 1
```

T. Villa et al., *The Unknown Component Problem: Theory and Applications*,
DOI 10.1007/978-0-387-68759-9_9, © Springer Science+Business Media, LLC 2012

```
-1 1
.names G16 G15 G9
0- 1
-0 1
.end
```

9.1 Extracting an Automaton from a Sequential Circuit

The operation `extract_aut` builds the automaton corresponding to the FSM extracted from a given sequential circuit, according to the construction introduced in Definition 3.1.7, by which the inputs and outputs of the FSM are merged to become the inputs of the automaton. In the example, the sequential circuit is given as a netlist in BLIF format.

```
read_blif S27.blif
extract_aut S27.aut
```

The result is the automaton *S27.aut* shown in Fig. 9.1.

The `S27.blif` file has 3 binary latches for a total of 8 possible states. However, the result of `extract_aut` indicates that only 6 states are reachable from the initial state 000. The octagonal node at the top indicates that it is the initial state. The output indicates that the automaton is incomplete with all 6 states being incomplete. Thus when we complete the automaton, there will be a transition from each state to the

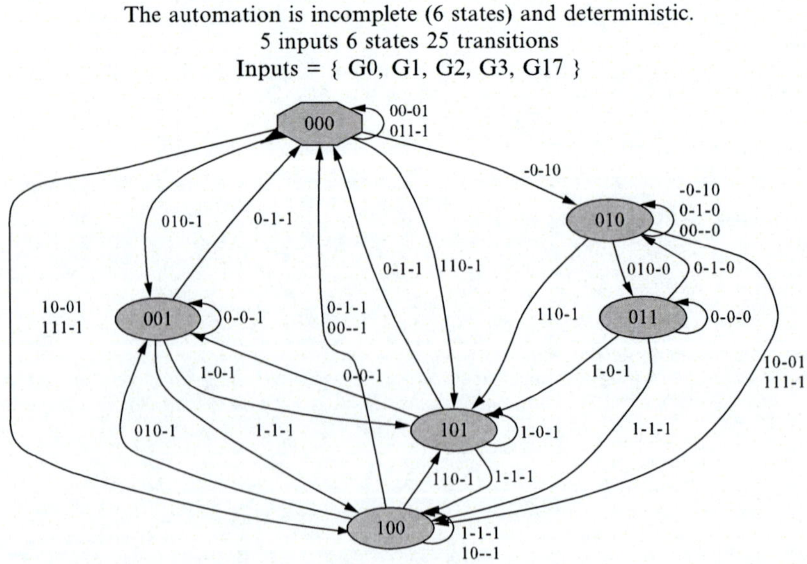

Fig. 9.1 Graphical output of BALM showing the automaton `s27.aut` obtained by extraction from the sequential circuit `S27.blif`

completion don't care state. The incompleteness is due to the output of the FSM becoming an input to the automaton, because by construction the inputs of the automaton are the cartesian product of the inputs and outputs of the FSM. The result is that a state of the automaton is not complete for those combinations of values of inputs and outputs of FSM such that at that state the FSM under those inputs does not produce those outputs.

9.2 Completing an Automaton

A state is called incomplete if under some input combinations no transition to a next state is defined. If an automaton has some incomplete states, it can be completed by adding one additional non-accepting state, called the don't care state.

```
complete S27.aut S27_cpl.aut
```

The result is the automaton *S27_cpl.aut* shown in Fig. 9.2.

Since this example came from an FSM, all states are accepting except the completion don't care state DC. Note that it is shaded differently to indicate a non-accepting state. Also note that there are 6 incoming transitions to DC (besides its universal self-loop), one from each state.

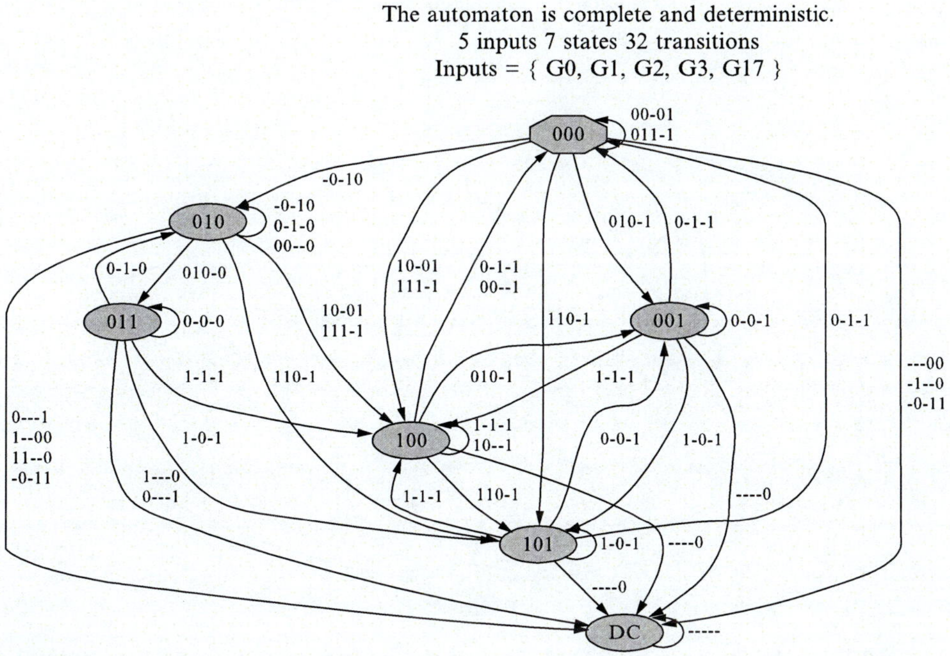

Fig. 9.2 Graphical output of BALM showing the automaton `S27_cpl.aut` obtained by completing the automaton `S27.aut`

9.3 Changing the Support

The support of an automaton is the list of its input variables. Its support can be changed by giving a new list. This one command can be used to hide variables or to insert new variables in the support, or to rearrange the order of the support list.

```
support G0,G2,G17 S27.aut S27_sup1.aut
```

The result is the automaton *S27_sup1.aut* shown in Fig. 9.3.

Initially, this had 5 variables in the support, but the support command caused two, $G1$ and $G3$, to be dropped. If we bring back these variables, their instances are replaced with -, indicating any value.

```
support G0,G1,G2,G3,G17 S27_sup1.aut S27_sup2.aut
```

The result is the automaton *S27_sup2.aut* shown in Fig. 9.4.

Notice that if a variable has more than two values one must declare the number of values between parentheses, e.g.,

```
support u(3),v,z(4) <file_in> <file_out>
```

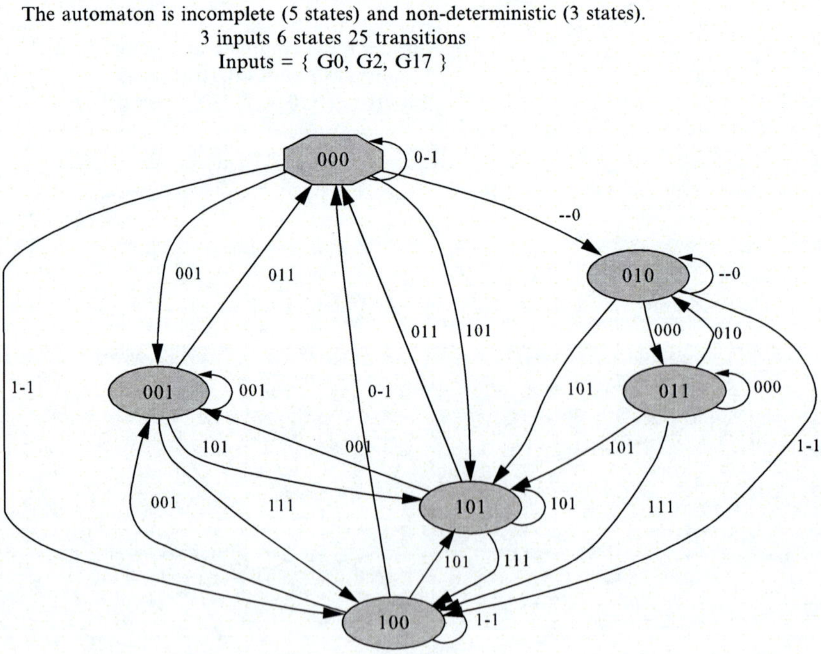

Fig. 9.3 Graphical output of BALM showing the automaton `s27_sup1.aut` obtained by changing the support of the automaton `S27.aut`

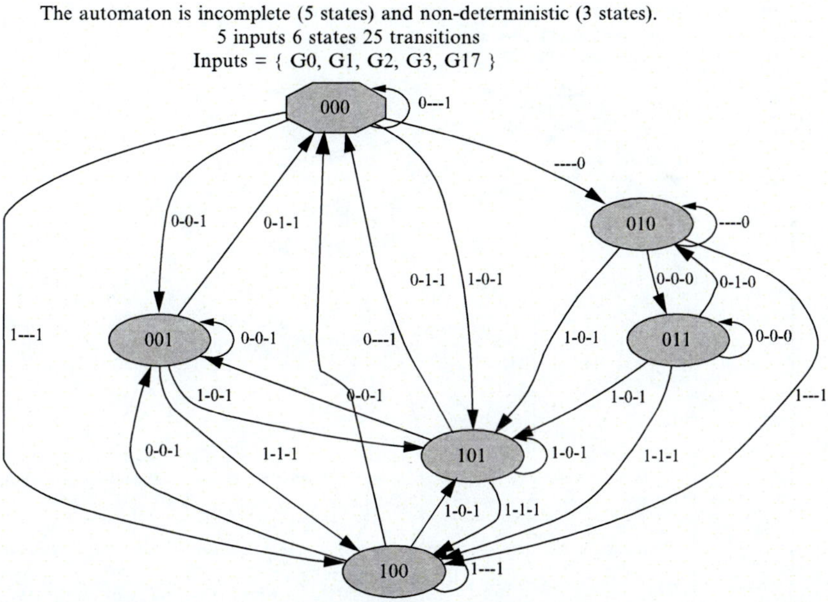

Fig. 9.4 Graphical output of BALM showing the automaton `s27_sup2.aut` obtained by changing the support of the automaton `S27_sup1.aut`

9.4 Determinizing

A non-deterministic automaton can be determinized using the subset construction in which subset states are created if a transition input exists, which can transit to at least two different states. Generally, the subset construction can lead to a set of new states, which potentially can be any subset of the set of original states, leading to a possible exponential number of states. In practice this almost never happens in practical examples, and in fact the new automaton may have a smaller number of states.

```
determinize -l S27_sup1.aut S27_sup1_det.aut
```

The result is the automaton *S27_sup1_det.aut* shown in Fig. 9.5.

Note that the names of the states have changed to subset names (the option `-l` keeps the original names to show which subset states were formed). For instance state 010_011 denotes a subset state composed of two states, 010 and 011. This came from the fact that `S27_sup1.aut` was non-deterministic, caused by having restricted its support to only 3 of the original 5 inputs. Note that there are still only a total of 6 states despite that a subset construction was done. This is because all possible subset states cannot be reached from the initial state.

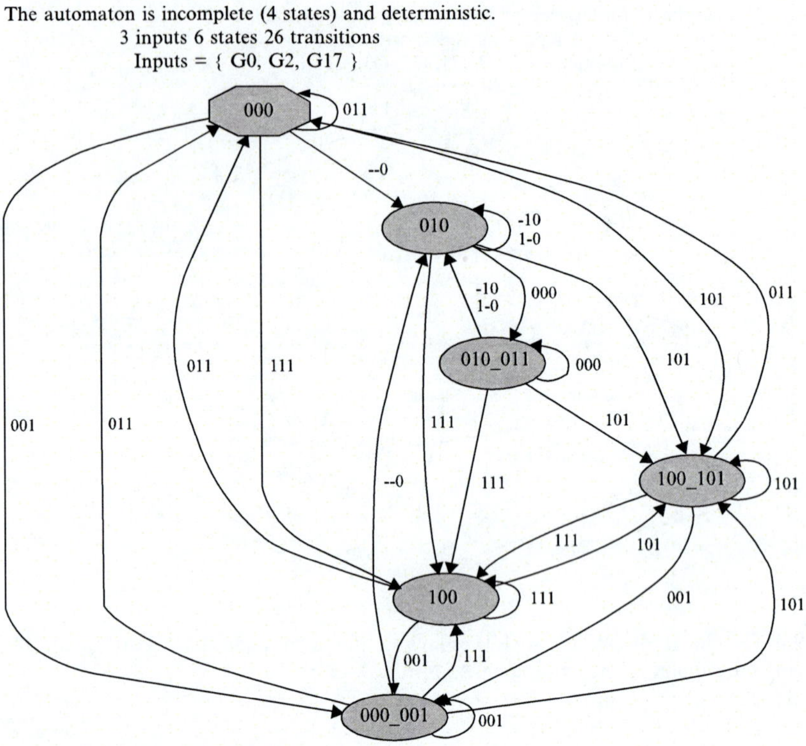

Fig. 9.5 Graphical output of BALM showing the automaton `S27_sup1_det.aut` obtained by determinizing the automaton `S27_sup1.aut`

9.5 Taking the Product of Two Automata

To illustrate the product of two automata, we consider two automata, `S27Fs.aut` shown in Fig. 9.6 and `S27as.aut` shown in Fig. 9.7. If the two automata have different supports, the user is responsible for lifting both automata to their common support. By an undocumented option, the product operation in BALM is defined to automatically lift both automata to their least common support. In the example below, the two automata start out with the same support.

```
product -l S27Fs.aut S27as.aut S27_prod.aut
```

The result is the automaton *S27_prod.aut* shown in Fig. 9.8.

Note that this automaton is isomorphic to `S27.aut`. This is because the two automata `S27Fs.aut` and `S27as.aut` were created by decomposing `S27.aut` into two sub-automata. When their product is formed, they recreate the original automaton. The state names of `S27Fs.aut` were changed from $0, 1$ to a, b respectively to help illustrate where the product states came from. Thus product state $b01$ came from state b in `S27Fs.aut` and from state 01 in `S27as.aut`.

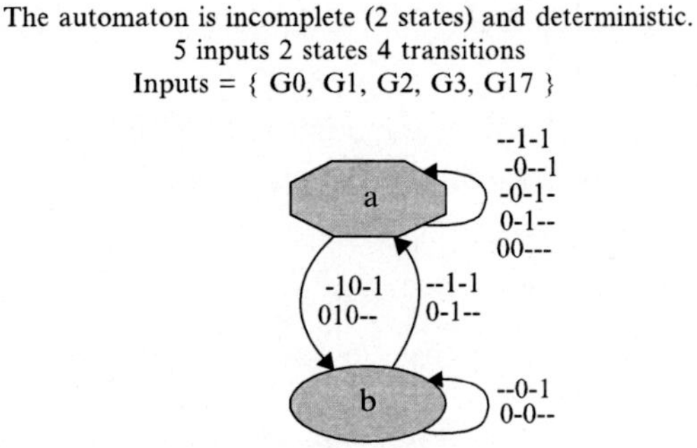

Fig. 9.6 Graphical output of BALM showing the automaton `S27Fs.aut`

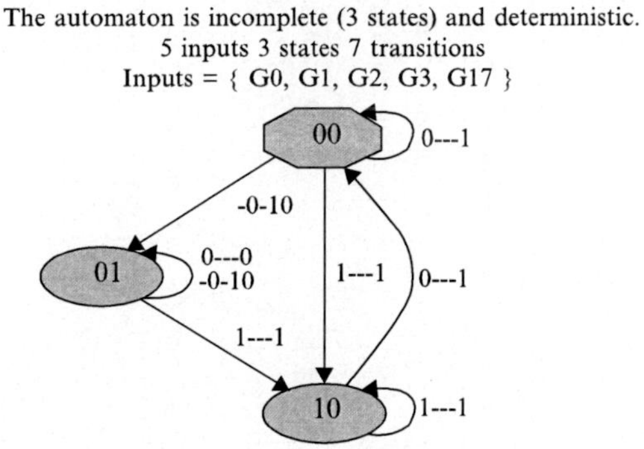

Fig. 9.7 Graphical output of BALM showing the automaton `S27as.aut`

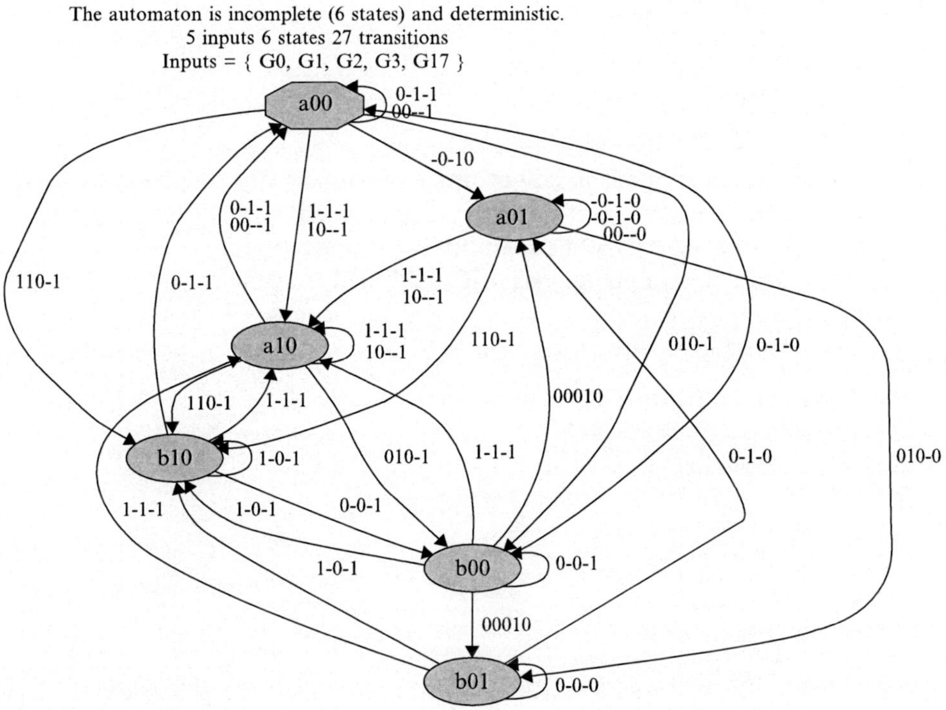

Fig. 9.8 Graphical output of BALM showing the automaton `S27_prod.aut` obtained by taking the product of the automata `S27Fs.aut` and `S27as.aut`

Problems

9.1. Consider the context FSM M_A and the specification FSM M_C whose descriptions in `kiss` format are, respectively, the following:

```
.i 2
.o 2
.s 1
.p 4
.r a

i1 v1 a a u1 o
i2 v1 a a u2 o
i1 v2 a a u2 o
i2 v2 a a u1 o
.end
```

and

```
.i 1
.o 1
.s 1
.p 2
.r b

i1 b b o
i2 b b o
.end
```

(a) Convert the descriptions in `kiss` format of M_A and M_C into their descriptions as automata in the `blif_mv` format, creating respectively the files `fig2ma.aut` for the context M_A and `fig2mc.aut` for the specification M_C.

Caveat: since not all commands of BALM handle one-state machines, a workaround is to rewrite the FSMs/automata with two equivalent states. In this case a solution for `fig2ma.aut` is:

```
.model fig2ma
.inputs i v u o
.outputs Acc

.mv CS, NS 2 a aa

.mv i 2 i1 i2
.mv v 2 v1 v2
.mv u 2 u1 u2
```

```
.latch NS CS
.reset CS
a

.table CS -> Acc
.default 1

.table i v u o CS -> NS
i1 v1 u1 0 a aa
i2 v1 u2 0 a aa
i1 v2 u2 0 a aa
i2 v2 u1 0 a aa

i1 v1 u1 0 aa a
i2 v1 u2 0 aa a
i1 v2 u2 0 aa a
i2 v2 u1 0 aa a

.end
```

and a solution for `fig2mc.aut` is:

```
.model fig2mc
.inputs i o
.outputs Acc


.mv i 2 i1 i2
.mv CS, NS 2 b bb

.latch NS CS
.reset CS
b

.table CS -> Acc
.default 1

.table i o CS -> NS

i1 0 b bb
i2 0 b bb

i1 0 bb b
i2 0 bb b

.end
```

(b) Solve the equation $M_A \bullet M_X \subseteq M_C$ by running the appropriate commands in BALM.

Hint. Run the following commands:

```
complement fig2mc.aut mc_comp.aut
product fig2ma.aut mc_comp.aut product.aut
support  v,u product.aut product_supp.aut
complement product_supp.aut x.aut
progressive -i 1 x.aut xfsm.aut
minimize xfsm.aut xfsm_min.aut
```

Analyze carefully the result of each step. How is the solution?

9.2. By running the appropriate commands in BALM, solve the same equation for the following context automaton `fig3ma`:

```
.model fig3ma
.inputs i v u o
.outputs Acc

.mv CS, NS 4 a b c d
.mv i 2 i1 i2
.mv v 2 v1 v2
.mv u 2 u1 u2
.mv o 2 o1 o2

.latch NS CS
.reset CS
a

.table CS -> Acc
.default 1

.table i v u o CS -> NS
i1 v1 u1 o1 a b
i2 v1 u1 o2 a b
i2 v2 u1 o2 a b

i2 v2 u2 o1 b b
i1 v1 u1 o2 b b
i1 v2 u1 o2 b b
i2 v1 u2 o1 b b

i1 v2 u2 o1 a c

i2 v1 u2 o2 c c
```

```
i1 v1 u1 o1 c d
i2 v2 u1 o2 c d
i1 v2 u2 o1 c d

i1 v1 u1 o2 d d
i2 v1 u1 o1 d d
i1 v2 u1 o2 d d
i2 v2 u1 o1 d d
.end
```

and specification automaton fig3mc:

```
.model fig3mc
.inputs i o
.outputs Acc

.mv CS, NS 4 a1 b2 c3 d3
.mv i 2 i1 i2
.mv o 2 o1 o2

.latch NS CS
.reset CS
a1

.table CS -> Acc
.default 1

.table i o CS -> NS
i1 o1 a1 b2
i2 o2 a1 b2

i1 o2 b2 b2
i2 o1 b2 b2

i1 o1 a1 c3

i1 o1 c3 d3

i1 o2 d3 d3
i2 o1 d3 d3
.end
```

Plot the state transition graph of the result xfsm_min.aut.

9.3. By running the appropriate commands in BALM, solve the same equation for the following context automaton `fig9ma`:

```
.model fig9ma
.inputs i v u o
.outputs Acc

.mv CS, NS 2 a b
.mv i 2 i1 i2
.mv v 3 v1 v2 v3
.mv u 2 u1 u2
.mv o 3 o1 o2 o3

.latch NS CS
.reset CS
a

.table CS -> Acc
.default 1

.table i v u o CS -> NS
i1 v1 u1 o1 a a
i1 v2 u2 o1 a a
i1 v3 u1 o1 a a
i2 v1 u2 o3 a a
i2 v2 u2 o1 a a
i1 v3 u2 o3 a a

i1 v2 u1 o2 a b
i2 v1 u1 o2 a b

i1 v2 u2 o1 b b
i2 v3 u2 o3 b b

i1 v3 u1 o1 b a
i2 v1 u2 o3 b a
i2 v2 u1 o3 b a
i1 v1 u1 o1 b a
.end
```

and specification automaton `fig9mc`:

```
.model fig9mc
.inputs i o
.outputs Acc
```

```
.mv CS, NS 2 a b
.mv i 2 i1 i2
.mv o 4 o1 o2 o3 o4

.latch NS CS
.reset CS
a

.table CS -> Acc
.default 1

.table i o CS -> NS
i1 o1 a b
i2 o1 a b
i2 o2 a b

i1 o1 b a
i2 o1 b a
i2 o2 b a
.end
```

9.4. Problem 3.3 asks to find the largest FSM solution of a given equation. Work out the same problem by using the suite of commands available in BALM to solve the equation.

By invoking the command `contain` in BALM, verify that the (language of the) largest solution computed before contains indeed the (language of the) FSM M_B that was used to obtain the specification $M_{A \bullet B} = M_A \bullet M_B$.

Part III
Application to Sequential Synthesis

In this part of the book, we look at the problem of sequential synthesis where a finite state machine is embedded in a larger FSM environment. The problem addressed is to find the set of all possible FSMs that can be used in place of the current FSM without changing the externally observed behavior of the FSM environment. This set is called the complete sequential flexibility, CSF, of the machine (with respect to its environment), and is analogous to the complete flexibility used in combinational synthesis. We will also show how to solve a number of examples in sequential synthesis using scripts of the BALM system, thereby reinforcing the understanding of its inner working from the user's point-of-view.

This part includes five chapters, of which the first four chapters address the problem of computing the flexibility in sequential networks (maximum or restricted flexibility); instead the last chapter discusses how to exploit the computed sequential flexibility by extracting a valid replacement that is advantageous with respect to a chosen cost function.

Chapter 10 reports some classical and new methods to compute restricted forms of sequential flexibility, as sequential don't cares; it surveys the techniques of Kim-Newborn, Wang-Brayton and Yevtushenko-Zharikova, illustrating some of them by examples solved with BALM.

Chapter 11 discusses how to compute the sequential flexibility in a netlist given in `BLIF-MV` or `BLIF` format, and then focus on a window, partitioning the netlist into two parts - all nodes inside the window and all nodes outside the window. The nodes inside the window can be viewed as a separate FSM and the nodes outside the window as its fixed environment.

Chapter 12 addresses the problem of resynthesizing the component FSMs of a network of FSMs; we will discuss both a global approach and a local (windowing) approach. It will turn out that sometimes it is more effective to solve a system of equations instead of a single equation; therefore we will introduce systems of equations over FSMs.

Chapter 13 discusses the use of simulation relations which make language solving easier for some topologies, but they do not always guarantee to compute the full flexibility. We show how such relations can be imposed by making some internal variables external.

Chapter 14 introduces the problem of selecting a "minimum" reduction or submachine of the largest solution. This is an introductory chapter about an open problem that does not have yet a satisfactory solution. Once we have the largest (or even a restricted) flexibility we need to select a deterministic behavior to implement in place of the original one. We would like to choose such a solution so that the final implemented logic is the best with respect to area, and/or delay, testability, power consumption and so on. So we need "minimizers" to select the best implementation; however, it is not obvious how to explore state-based representations with cost functions defined at the gate-level, and this makes quite difficult the task. For instance, a common criterion of selecting a deterministic behavior is to find a state-based representation with a minimum number of states, however this does not track necessarily well with a minimum area at the gate-level.

Chapter 10
Computation of Flexibility in Sequential Networks

There is a long history of resynthesizing an FSM, given its surrounding environment. Much of the work was modeled after results for combinational networks. Thus input sequential don't cares and output sequential don't cares were defined in analogy to satisfiability and observability don't cares. For example, input sequential don't cares were defined as input sequences that can never happen at a state because of the FSM input environment. An elegant theory was provided by Kim and Newborn for treating the case of a cascade of two FSMs. This is discussed in Sect. 10.1 as well as an extension by Wang and Brayton in Sect. 10.2. These results provide reasonable computational procedures and can be used for resynthesis of an FSM. However, attempts at extending to output sequential don't cares became overly complicated and were unsuccessful. It was surprising then that Watanabe came up with a computation for the full flexibility for an FSM embedded in an environment, which captures all input and output sequential don't cares. This was called the "E-machine" and was constructed by a somewhat complicated construction. It became clear later that this construction essentially modeled the subset construction. Now we know that this full flexibility embodied by the E-machine is simply the largest FSM solution obtained by language solving, and a simpler construction is the one given in this book and discussed in this chapter in more detail.

10.1 The Kim–Newborn's Procedure

J. Kim and M.M. Newborn [69] devised a procedure to compute an incompletely specified machine which takes account of all input sequences which can never occur in a cascade connection of two FSMs. The topology considered is shown in Fig. 10.1, where M_A and M_B are FSMs.

We review the procedure:

1. Construct an NFA A' to accept the u-language produced by machine M_A. This can be achieved by removing the input part in the STG of M_A, and considering

T. Villa et al., *The Unknown Component Problem: Theory and Applications*, DOI 10.1007/978-0-387-68759-9_10, © Springer Science+Business Media, LLC 2012

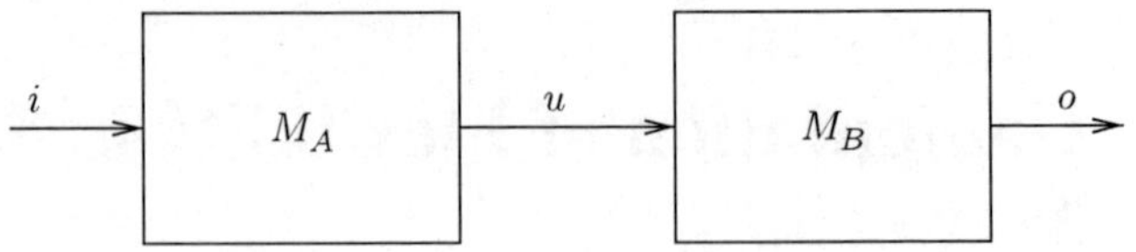

Fig. 10.1 Topology for Kim and Newborn procedure

every state of M_A as a final state. For a state s, if there are output symbols not emitted from it, a transition with these symbols is inserted from s to the dead state d. The dead state d is the only non-accepting state. Thus A' is completely specified but nondeterministic in general.

2. Convert A' to a minimized completely specified DFA A. This can be done by using the subset construction and state minimization for DFA A. Note that the efficient state minimization method *minimize* in BALM can be used, since the subset construction produces a deterministic machine.
3. A modified machine M'_B is constructed as follows: construct $M_B \times A$ and then redirect to a DNC state all transitions to a state that contains the dead state d in its subset. M'_B is an incompletely specified FSM.

We assume that M_A and M_B are given in BLIF-MV format as `MA.mv` and `MB.mv`, respectively. We can translate these steps into the equivalent BALM commands by the following script.

```
K&N script
read_blif_mv MA.mv
extract_aut MA.aut           \# convert into an automaton
support u MA.aut MAs.aut \# remove input
complete MAs.aut MAsc.aut \# add the dead state
determinize MAsc.aut MAd.aut
minimize MAd.aut MAd_min.aut
read_blif_mv MB.mv
extract_aut MB.aut           \# convert into an automaton
product MB.aut MAd_min.aut MBi.aut \# MB x A
prefix MBi.aut MBifsm.aut \# make prefix-closed
```

We illustrate this with a small example.

```
Machine MA.mv
.model MA
.inputs a b c d e
.outputs x y
.latch R1 r1
.reset r1
0
.latch R2 r2
.reset r2
0
.table c d x1
```

```
.default 0
1 - 1
- 1 1
.table r1 r2 x2
.default 0
1 - 1
- 1 1
.table x1 x2 x
.default 0
1 1 1
.table a b y1
.default 0
1 - 1
- 1 1
.table y1 e y
.default 0
1 1 1
.table a R1
.default 0
1 1
.table b R2
.default 0
1 1
.end

Machine MB.mv
.model MB
.inputs x y
.outputs o
.latch R1 r1
.reset r1
0
.latch R2 r2
.reset r2
0
.latch R3 r3
.reset r3
0
.table x y r1 r2 r3 o
.default 0
- 1 - - - 1
1 - - - 1 1
1 - 1 1 - 1
.table y r1 r2 r3 R1
.default 0
```

```
1 1 1 - 1
1 - - 1 1
.table x r2 R2
.default 0
1 1 1
.table x y r3 R3
.default 0
1 - 1 1
- 1 - 1
.end
```

We adapt the K&N script to this example:

```
K&N script - example MA.mv MB.mv
read_blif_mv MA.mv
extract_aut MA.aut
support x,y MA.aut MAs.aut
complete MAs.aut MAsc.aut
determinize MAsc.aut MAd.aut
minimize MAd.aut MAd_min.aut
read_blif_mv MB.mv
extract_aut MB.aut
product MB.aut MAd_min.aut MBi.aut
prefix MBi.aut MBifsm.aut
```

We obtain the automaton shown in Fig. 10.2, which is reduced to two states by state minimization with STAMINA.

In contrast, we can solve for the largest FSM solution of the component M_B of this example. We take the specification in `spec.mv` to be the FSM composed of $M_A \times M_B$ with inputs a, b, c, d, e and output o. We use the following script:

```
Largest solution script - example spec.mv MA.mv
read_blif_mv spec.mv
extract_aut spec.aut
complement spec.aut specc.aut
read_blif_mv MA.mv
extract_aut MA.aut
product MA.aut specc.aut p.aut
support x,y,o p.aut ps.aut
complement ps.aut x.aut
progressive -i 2 x.aut xfsm.aut
minimize xfsm.aut temp.aut
prefix temp.aut xfsm_min.aut
```

We obtain the solution `xfsm_min.aut` shown in Fig. 10.3, which is again reduced to two states by state minimization with STAMINA.

One can check that `MBifsm.aut` is contained in `xfsm_min.aut` using the command `contain MBifsm.aut xfsm_min.aut`.

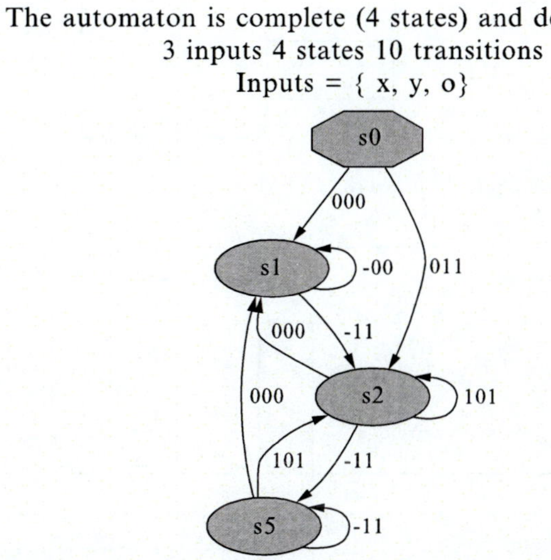

Fig. 10.2 Graphical output of BALM showing the automaton `MBifsm.aut` generated by the script `K&N script - example MA.mv MB.mv`. The first two variables are the internal signals x, y, whereas the third variables is the external output o

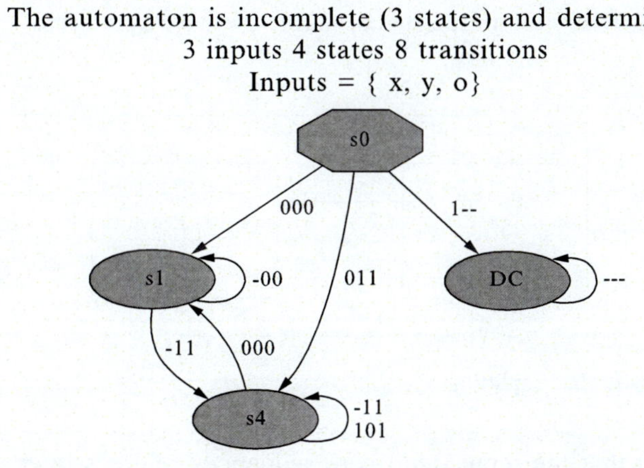

Fig. 10.3 Graphical output of BALM showing the automaton `xfsm_min.aut` generated by the script `Largest solution script - example spec.mv MB.mv` that computes the largest solution

10.2 The Wang–Brayton's Procedure

H.-Y. Wang and R. Brayton [139] extended the method of Kim and Newborn to include the case where M_B had a feedback connection to M_A. This configuration is shown in Fig. 10.4. This theory also applies in the case where the output o is produced by M_A.

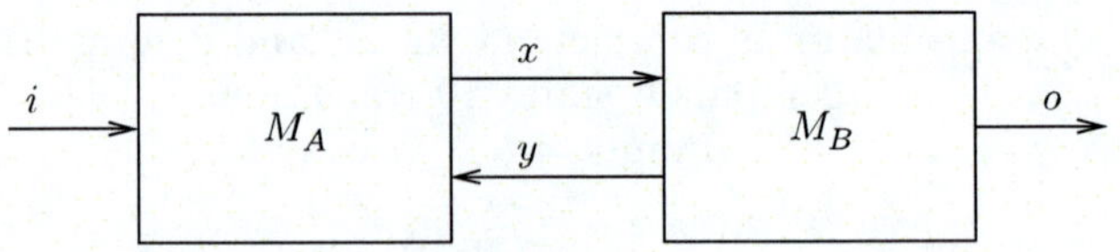

Fig. 10.4 A two-way communication network of FSMs

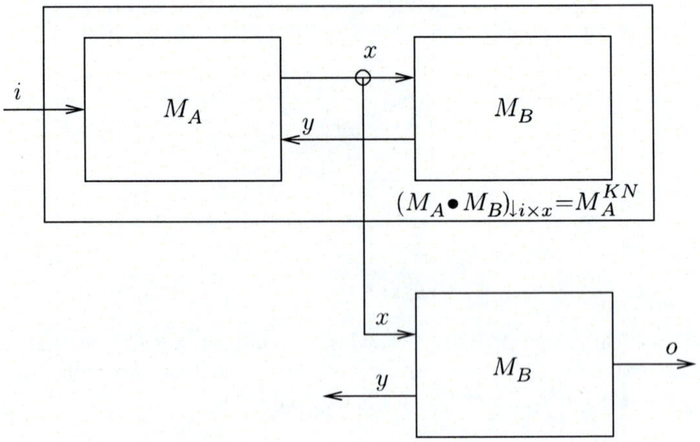

Fig. 10.5 Wang–Brayton's construction to extend Kim–Newborn's procedure

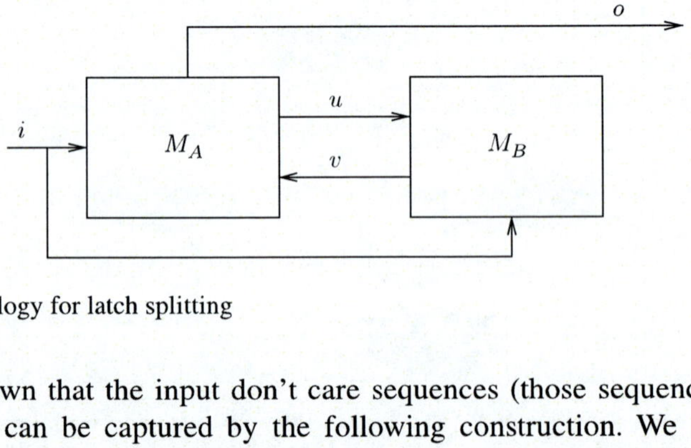

Fig. 10.6 Topology for latch splitting

It was shown that the input don't care sequences (those sequences which can never occur) can be captured by the following construction. We derive $(M_A \bullet M_B)_{\downarrow i \times x}$, the composition FSM of M_A and M_B with input i to M_A and output x to M_B as shown in Fig. 10.5. Then the Kim–Newborn construction is applied where M_A (in Kim–Newborn's procedure) is replaced by $(M_A \bullet M_B)_{\downarrow i \times x} = M_A^{KN}$ (in Wang–Brayton's procedure).

The argument for the correctness of this is that the composition will produce the same sequences of x is as in Fig. 10.4. Since the configuration is now the same as in Sect. 10.1, the Kim–Newborn procedure can be used. We will illustrate this on an example and compare the incompletely specified FSM produced in this way with the largest FSM solution produced by language solving.

We use an example which is produced by latch splitting. The latch splitting topology is shown in Fig. 10.6, where M_B becomes the unknown component.

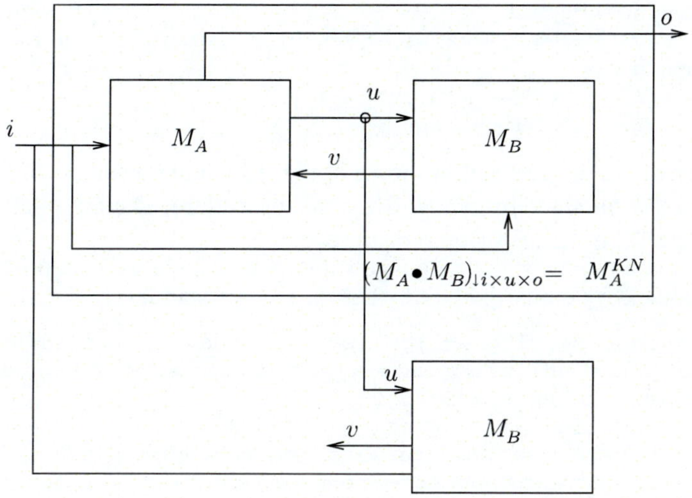

Fig. 10.7 Topology for Wang–Brayton's construction applied to latch-splitting example

This is converted into the Kim–Newborn topology shown in Fig. 10.7.

We use the example `s386.blif` from the iscas89 benchmark suite and do latch splitting on this. The inputs i are `v6,v5,v4,v3,v2,v1,v0`. The u vector, `v9,v8,v7` and the v vector, `v12,v11,v10`, are produced by the `latch_split` command. The BALM commands (with comments) executed are

```
W&B-K&N script - example s386.blif
read_blif s386.blif
latch_split 0-2
 \# produces files s386f.blif and s386a.blif
latch_expose
 \# exposes all latch outputs v7 - v12 of s386.blif
extract_aut MA.aut
 \# since s386.blif is already the product MA x MB, this is
 \# MA x MB and will become the Kim--Newborn abstract driving
 \# machine MA^RN for MB (it is the composition of s386a.blif
 \# and s386f.blif which is just s386.blif)
support v6,v5,v4,v3,v2,v1,v0,v9,v8,v7,v12,v11,v10 MA.aut
    MAs.aut
 \# restricts support to i,u,v
complete MAs.aut MAsc.aut
determinize MAsc.aut MAd.aut
minimize MAd.aut MAd_min.aut
 \# this is the abstract driving machine MA^KN of Kim--Newborn
 \# for MB
read_blif s386a.blif
 \# this is MB
latch_expose
 \# we need to expose the v variables which are the latch
 \# outputs of MB
 \# forces the v variables of MA x MB to agree with the v
 \# variables of MB
```

```
extract_aut MB.aut
support v6,v5,v4,v3,v2,v1,v0,v9,v8,v7,v12,v11,v10 MB.aut
    MBs.aut
product MBs.aut MAd_min.aut MBi.aut
prefix MBi.aut MBifsm.aut
```

This produces the automaton shown in Fig. 10.8 that represents the incompletely specified FSM obtained by Wang–Brayton's procedure. The input labels have been left off the STG because it would be unreadable.

To compare this with the language solving approach, we execute the latch split script produced by the BALM `latch_split` command:

```
\# Language solving script (partitioned) generated by MVSIS
\# for latch splitting of sequential network s386.blif
\# on Wed Jun 06 09:02:05 2007
\# Command line was: latch_split  0-2.
\# We use the previous results generated by latch-splitting
\# s386.blif, which created the environment automaton
\# s386f.blif

echo Solving the language equation ...
solve_fsm_equ s386f.blif s386.blif v6,v5,v4,v3,v2,v1,v0,v9,
v8,v7 v12,v11,v10 s386xs.aut
print_stats_aut s386xs.aut
```

Executing `print_stats_aut` gives the following output

```
Solving the language equation ...
Progressive =    0.00 sec
"csf":incomplete(14 st), deterministic, non-progressive(14 st),
and non-Moore(14 st).
13 inputs(13 FSM inputs)  15 states(15 accepting)  59 trans
Inputs = { v6,v5,v4,v3,v2,v1,v0,v9,v8,v7,v12,v11,v10 }
```

We now check if the language of the automaton `MBifsm.aut`, produced by the W&B-K&N procedure, is contained in the most general FSM solution, `s386xs.aut`:

```
balm > contain s386xs.aut MBifsm.aut
Warning: Automaton 1 is completed before checking.
Warning: Automaton 2 is completed before checking.
The behavior of automaton 1 contains the behavior of
    automaton 2.
```

and fortunately it is.

The next steps would be to use the incompleteness of `MBifsm.aut` to state minimize it as an incompletely specified FSM. The program STAMINA could be used. In contrast, the NDFSM `s386xs.aut`, produced by language solving, needs to be minimized but there are no good general procedures for this at the moment. The operator `dcmin` can be used and in this case it does produce an NDFSM automaton with 6 states (from 15 states), but the original machine for this part, `s386a.blif`, has only 5 reachable states. We did not try STAMINA on `MBifsm.aut` to see what kind of minimization can be done.

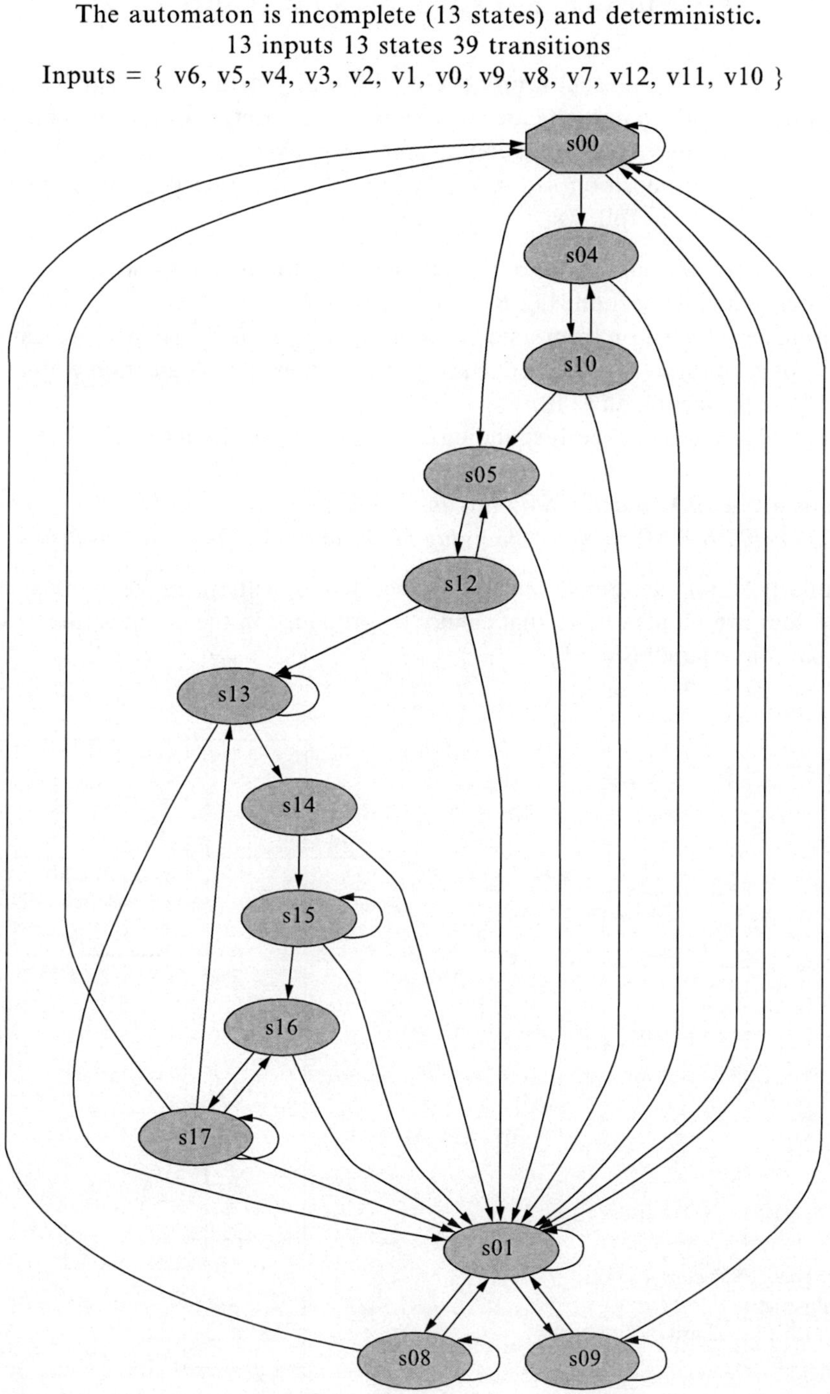

Fig. 10.8 Graphical output of BALM showing the automaton `MBifsm.aut` generated by the script `W&B-K&N script - example s386.blif`. The automaton represents the incompletely specified FSM created by Wang–Brayton's procedure

10.3 The Yevtushenko–Zharikova's Procedure

In this section, we present a procedure due to N. Yevtushenko and S. Zharikova to compute sequential don't cares that capture a restricted form of flexibility, to be applied for the resynthesis of a component in a network of FSMs (see [151, 153]).

Given the composition $M_A \bullet M_B$ in Fig. 10.9, the procedure by Yevtushenko–Zharikova works as follows:

1. Transform M_A and M_B into the corresponding automata A and B.
2. Derive the intersection $A \cap B_{\uparrow I \times O}$.
3. Replace a transition from state b_k under input u_j in B by a don't care transition [1] if for each state $(a, b_k) \in A \cap B_{\uparrow I \times O}, a \in A$, there is no transition with an action whose projection on U is u_j.
4. Let M'_B be the final corresponding incompletely specified FSM.

Proposition 10.1. *Each FSM that is a reduction of FSM M'_B can replace the component FSM M_B without changing the behavior of the overall system.*

Similarly, don't care transitions can replace care transitions in the component FSM M_A; they are inputs of M_A that cannot be produced in the composition as outputs by the component FSM M_B.

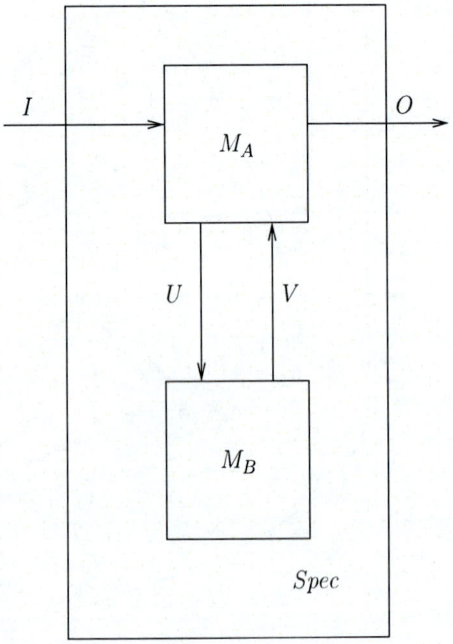

Fig. 10.9 FSM composition to illustrate Yevtushenko–Zharikova's procedure

[1] A don't care transition goes to a designated DNC state and issues any output; at the state DNC under any input there is a self-loop that issues any output.

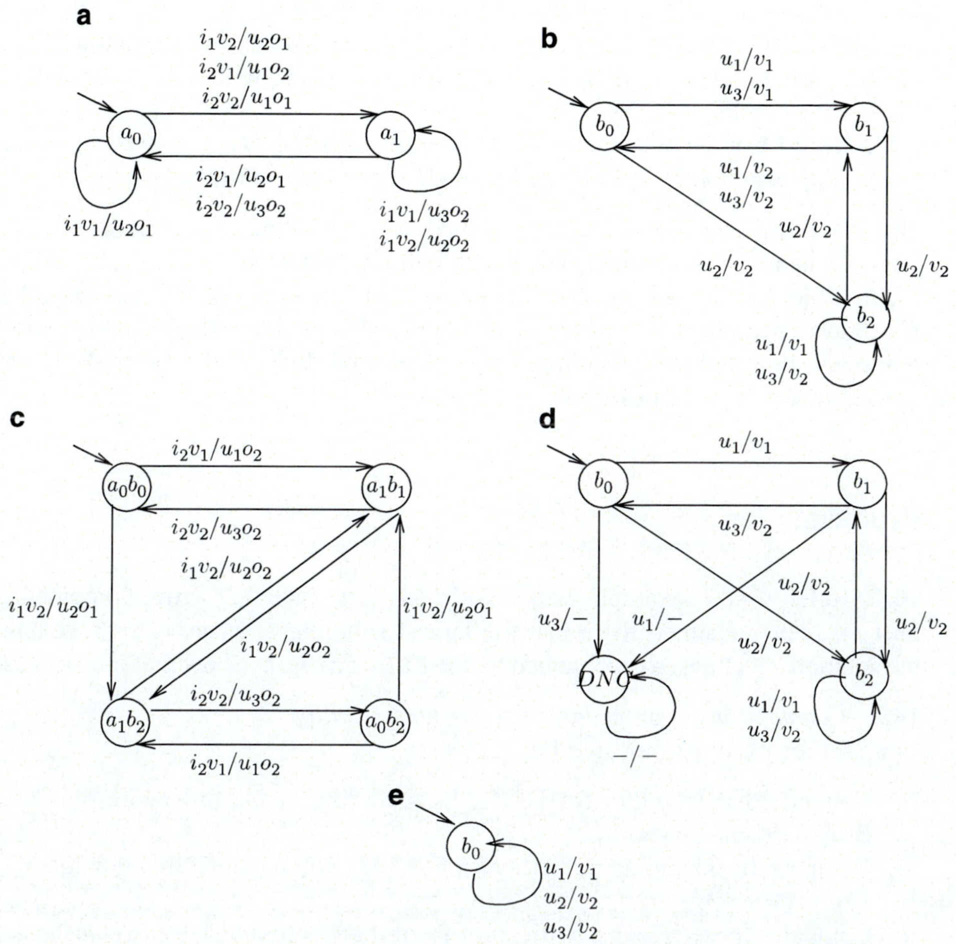

Fig. 10.10 FSMs with reference to the topology in Fig. 10.9. **(a)** FSM M_A; **(b)** FSM M_B; **(c)** Intersection $M_A \cap M_{B_{\uparrow I \times O}}$; **(d)** Incompletely specified FSM M'_B; **(e)** FSM M'_{B_R} that is a reduction of M'_B

Example 10.2. Consider the transition diagrams of FSMs M_A and M_B in Fig. 10.10a,b. Derive the intersection $M_A \cap M_{B_{\uparrow I \times O}}$, which is shown in Fig. 10.10c; the states of the intersection are the pairs $\{a_0b_0, a_1b_1, a_0b_2, a_1b_2\}$.

For each state of the FSM M_A, determine the don't care inputs over alphabet U. The don't care input of state a_0 is given by the set $DC_{a_0} = \{u_3\}$, since $\{u_3\}$ cannot be produced at state a_0. In the same way, we obtain $DC_{a_1} = \{u_1\}$.

Now check all the states of the FSM M_B. In the intersection, state b_0 is combined only with state a_0. Since $DC_{a_0} = \{u_3\}$, the transition in M_B from state b_0 under input u_3 can be replaced by a don't care transition. State b_1 is combined only with state a_1; thus, the transition from state b_1 under input u_1 can be replaced by a don't care transition. Finally, state b_2 is combined with state a_0 as well as with state a_1, and so there are no don't care transitions from state b_2, since $DC_{a_0} \cap DC_{a_1} = \{u_3\} \cap \{u_1\} = \emptyset$. In other words, all missing transitions are added as don't care transitions, i.e., they are directed to the DNC state. The obtained incompletely specified FSM M'_B is shown in Fig. 10.10d.

Each reduction of M'_B can replace the component FSM M_B, e.g., we can replace M_B by the reduction FSM M'_{B_R} with a single state that is portrayed in Fig. 10.10e.

This procedure has the advantage that it works directly on machine M_B (or M_A) and can obtain a result that is no worse than the original machine.

An FSM that is a reduction of the ISFSM obtained by the Yevtushenko–Zharikova's procedure is a reduction of the ISFSM obtained by the Wang–Brayton's procedure, but the vice versa does not always hold. Indeed, the Wang–Brayton's procedure allows more flexibility.

Problems

10.1. Refer to the example with FSMs `MA.aut` and `MB.aut` computed in Sect. 10.1. It is claimed there that the largest solution `xfsm_min.aut` contains the solution `MBifsm.aut` obtained by the `K&N script`. Show that it is the case.

10.2. *Comparisons of procedures to compute flexibility*
Consider the example shown in Fig. 10.10.

(a) Compute the flexibility for FSM M_B according to the procedure of Wang–Brayton/Kim–Newborn.
(b) Compute the maximum flexibility for FSM M_B solving the FSM equation $M_A \bullet M_X = M_A \bullet M_B$.
(c) Compare these results with the flexibility computed by Yevtushenko–Zharikova's procedure.

Chapter 11
Computation of Sequential Flexibility in Netlists by Windowing

11.1 Complete Sequential Flexibility in a Window

A challenging problem is to take a larger FSM given as a netlist, e.g., in `BLIF-MV` or `BLIF` format, and then focus on a window, partitioning the netlist into two parts – all nodes inside the window and all nodes outside the window. The nodes inside the window can be viewed as a separate FSM and the nodes outside the window as its fixed environment. The nodes on the boundary of the window are sometimes called a cut – they separate the inside from the outside. The nodes which fanout from the inside to the outside are the POs of the window and the nodes which fanout from the outside to the inside are the PIs of the window FSM. A similar situation occurs when the initial netlist is given as a hierarchy of FSMs. In this case the boundaries between the FSMs are given by the hierarchy structure.

The netlist inside the window will be considered as the unknown FSM component; the remaining nodes will constitute the fixed part. The specification will be the entire netlist. The largest FSM solution of the unknown component problem is sometimes referred to as the complete sequential flexibility (CSF) of the window. It contains all deterministic FSMs, which can be placed in the window and connected to the fixed part such that the overall composed behavior is the same as the original netlist. Note that unlike some problems, this application comes with a particular known solution, which is the original netlist in the window. The challenge of this application is to come up with a replacement netlist which is better, in some sense. Even though the particular known solution must be contained in the largest FSM solution, it is surprisingly difficult in many cases to find a sub-machine that is equal or better than the original known solution. One reason is that the largest FSM solution is usually a relatively very large machine and has so much behavior that is it hard to find a small sub-solution.

In practical applications, the original netlist is too large, with too many latches for the techniques in this book to be applied. To overcome this, a typical fix is to use a two-window approach, where the first step is to find a sub-netlist which is the target of resynthesis, in order to obtain some improvement over the original netlist inside

T. Villa et al., *The Unknown Component Problem: Theory and Applications*,
DOI 10.1007/978-0-387-68759-9_11,

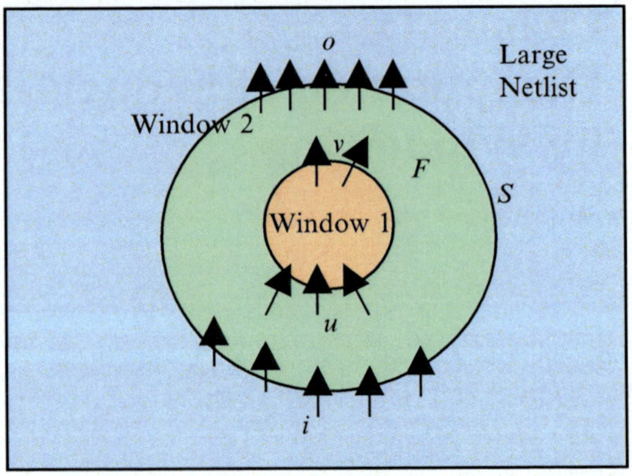

Fig. 11.1 A window and its abstracted environment, S

the window, if possible. Then a second window is found which contains the first window. The second window is chosen to include as much information about the flexibility allowed by the environment without making the second window too large. An example of choosing the second window would be to include all nodes in the transitive fanout of up to k levels and all nodes in the transitive fanin up to m levels. Even this might be too large, so some methods for trimming this second window can be used. An efficient method for choosing the second window is proposed in Sect. 11.2. It is modeled after similar constructions for combinational networks.

Once the two windows have been chosen, the larger one is taken as the specification and the smaller one as the unknown component. Note that the efficient method of solution of Chap. 7 can be used for obtaining the largest FSM solution for the smaller window, because both the specification and fixed part are given as deterministic netlists. Suppose that the FSM solution has been obtained and that a small implementation is chosen to replace the first window, then the overall behavior of the larger window is unchanged according to the theory of language solving. Hence, the overall behavior of the netlist is also unchanged.

We highlight that in all steps of windowing, from choosing the second window of the unknown component to selecting a replacement from the largest solution and then optimizing the resulting sequential logic, care must be observed not to introduce combinational cycles. However we will not discuss the issue in this chapter, and we leave it as a caveat in the background.

This idea is used for minimizing the original large netlist by iterating this procedure and choosing another window for minimization, solving and reimplementing the network inside the window. This is iterated over the entire network. By limiting the sizes of the windows chosen, the computations can be made practical and reasonably efficient.

In this chapter, we discuss the methods for choosing the two windows and the implications that their sizes have on the solution process.

The general setting is shown in Fig. 11.1, where *Window 1* denotes the sub-network, X, to be optimized and *Window 2* denotes the specification, S, to be used in the language solving instance. The sub-network $S \setminus X$ (S without X) represents the fixed part, F, of the language solving problem.

Let `F.mv` describe the netlist of the FSM of the fixed part and `S.mv` the netlist of the FSM of the specification, which is the netlist of the second window. As shown in Fig. 11.1, the variables i and v are inputs to F and the variables u and o are the outputs of F. Using BALM, there are two ways to find the CSF of the netlist of the inner window. The first way is to run the sequence of BALM procedures:

```
read_blif_mv S.mv
extract_aut S.aut
complement S.aut Sc.aut
read_blif_mv F.mv
extract_aut F.aut
product F.aut Sc.aut FxSc.aut
support u1,u2,u3,v1,v2 FxSc.aut FxScs.aut
complement FxScs.aut x.aut
progressive -i 3 x.aut xfsm.aut
```

Here, for the sake of example, we assume that `u1,u2,u3` and `v1,v2` are the binary inputs and outputs of the unknown component. If they are non-binary variables, then the sizes of their domains would need to be given in the command `support`, e.g.,

```
support u1(3),u2(3),u3(3),v1(4),v2(5) FxSc.aut
  FxScs.aut.
```

The second way of solving for the CSF is to use the command `solve_fsm_equ`, as in:

```
solve_fsm_equ F.mv S.mv u1,u2,u3 v1,v2 xfsm.aut
```

where the two lists for u and v are comma-separated lists denoting the inputs and outputs. Even if the u, v variables are non-binary, it is not necessary to specify their domains since this information will be extracted from the files. These two procedures lead to identical automata, `xfsm.aut`.

11.2 Selecting a Window by Latch Splitting

One way that BALM allows for selecting a window is to take an entire FSM and use "latch splitting" on it. The command is called `latch_split`, which specifies which latches are to be put in each part. This separates the latches of the FSM into two parts, one part is in the fixed part F and the other part is in the known (original) solution for the unknown component. The topology is shown in Fig. 11.2, where the FSM S may be part of a larger netlist.

As shown in the figure, the outputs v of X have the dimension of the number of latches that were included in X. These may have nothing to do with the final states of X, since X is the unknown component to be solved for, and can have a completely different set of states and latches that were originally in the X part.

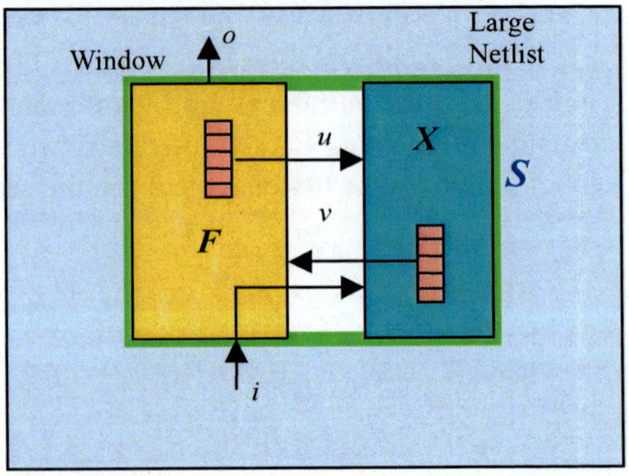

Fig. 11.2 Topology of the result of latch splitting

The known solution for X is called A and in this case the outputs v are the outputs of the latches. Note that because the outputs u of F are from latches, the configuration in Fig. 11.2 can have no combinational loops. Generally, S is a Mealy machine, so the input i needs to be an input of X, as shown.

The method of latch splitting will be illustrated with an example in more detail in Sect. 13.3; there we will set up some extra correspondences between signals that will force a simulation relation between state signals. The use of simulation relations is the topic of Chap. 13.

Another way to select a window is to focus on a particular section of a netlist for optimization. Around this section we can construct a surrounding window which will serve as the specification for the section. How to compute a reasonable surrounding window is the next topic.

11.3 Efficient Window Selection

Windowing is a method of limiting the scope of logic synthesis to work only on a small portion of a Boolean network. This method is indispensable for scalability when working with large Boolean networks arising in industrial applications.

11.3.1 Efficient Window Selection for Combinational Networks

The material in this section is adapted from [96], where windowing is used to compute don't cares in large combinational networks.

Definition 11.3.1. Two non-overlapping subsets of nodes, the leaf set and the root set, are in a **leaf/root relation** if every path from the PIs to any node in the root set passes through some node in the leaf set.

```
(nodeset, nodeset) Window(node N, int nFanins, int nFanouts) {
    nodeset I1 = CollectNodesTFI({N}, nFanins);
    nodeset O1 = CollectNodesTFO({N}, nFanouts);
    nodeset I2 = CollectNodesTFI(O1, nFanins + nFanouts);
    nodeset O2 = CollectNodesTFO(I1, nFanins + nFanouts);
    nodeset S = I2 ∩ O2;
    nodeset L = CollectLeaves(S);
    nodeset R = CollectRoots(S);
    return (L, R);
}
```

Fig. 11.3 Computation of a window for a node

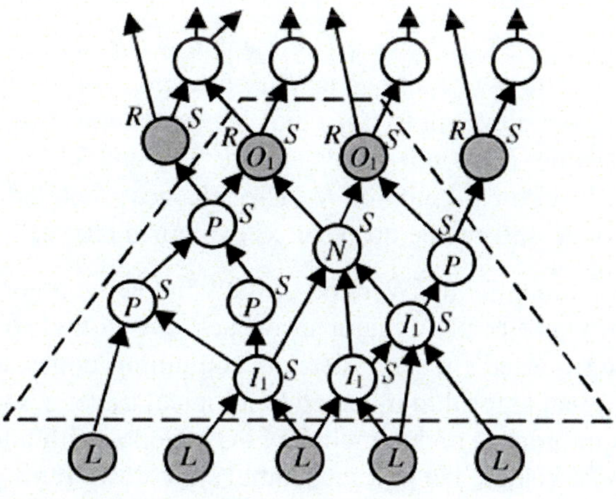

Fig. 11.4 Example of the 1×1 window of node N

Definition 11.3.2. Given two subsets in a leaf/root relation, its **window** is the subset of nodes of the network containing the root set together with all nodes on paths between the leaf set and the root set. The nodes in the leaf set are not included in the window.

Definition 11.3.3. A **path** connecting a pair of nodes is **distance-**k if it spans exactly k edges between the pair. Two **nodes** are **distance-**k from each other if the shortest path between them is distance-k.

The pseudo-code in Fig. 11.3 and the example in Fig. 11.4 describe the flow of a window construction algorithm. Procedure **Window** takes a node and two integers defining the number of logic levels on the fanin/fanout sides of the node to be included in the window. It returns the leaf set and the root set of the window.

The procedure **CollectNodesTFI** takes a set S of nodes and an integer $m \geq 0$, and returns a set of nodes on the fanin side that are distance-m or less from the nodes in S. An efficient implementation of this procedure for small m (for most applications, $m \leq 10$) iterates through the nodes that are distance-k ($0 \leq k \leq m$)

from the given set. The distance-0 nodes are the original nodes. The distance-$(k+1)$ nodes are found by collecting the fanins of the distance-k nodes not visited before. The procedure **CollectNodesTFO** is similar.

Procedures **CollectLeaves** and **CollectRoots** take the set of the windows internal nodes and determine the leaves and roots of this window. The leaves are the nodes that do not belong to the given set but are fanins of at least one of the nodes in the set. The roots are the nodes that belong to the given set and are also fanins of at least one node not in the set. Note that some of the roots thus computed are not in the TFO cone of the original node, for which the window is being computed, and therefore can be dropped without violating the definition of the window and undermining the usefulness of the window for logic synthesis operations dealing with the node.

We refer to the window constructed for a node by including n TFI logic levels and m TFO logic levels as an $n \times m$ window.

Example 11.1. Figure 11.4 shows a 1×1 window for node N in a network. The nodes labeled I_1, O_1, S, L, and R are in correspondence with the pseudo-code in Fig. 11.3. The windows roots (top) and leaves (bottom) are shaded. Note that the nodes labeled by P do not belong to the TFI and TFO of N, but represent reconvergent paths in the vicinity of N. The left-most root and right-most root are not in the TFO of N and can be dropped, as explained above.

Since implementing the first version of the windowing algorithm in [96], it was applied in several projects. A major drawback was found to be non-robustness when windowing is used for large designs containing nodes with more than 100 fanouts. The original algorithm involved topological traversals of the network from the window PIs in order to find the window POs. Nodes with multiple fanouts, each of which had to be visited, led to a substantial slow-down during this traversal. The problem was aggravated by the fact that multiple-fanout nodes were involved in many windows and, therefore, had to be traversed many times.

This led to the following modification. The original algorithm first detected the window PIs, then the window POs. The current algorithm does the opposite: it performs a shallow topological traversal to detect the window POs followed by a deeper reverse-topological traversal from the POs to find the window PIs. The topological traversal is performed with a fanout limit set to 10. The limit stops the traversal at multiple-fanout nodes and declares them as window POs because they are unlikely to yield any observability don't cares (due to many outgoing paths).

Another important improvement is that only those window POs are computed that have reconvergence involving the pivot node and the window PIs. The POs without reconvergence should not be included in the window because they do not contribute don't cares.

Once the window for a node is constructed, it is considered as the network, for the purpose of don't care computation or other logic optimizations. For this reason, algorithms for logic optimization can be discussed in the context of a sufficiently small network.

Part IV
More Applications of the Unknown Component Problem

The unknown component problem can be applied in several domains, such as: logic synthesis; model matching; testing; game theory; cryptography (FSM-based cryptosystems). Logic synthesis has been thoroughly discussed in Part III. In this part we consider some more applications: supervisory control, testing, game theory, synthesis with respect to ω-specifications.

Chapter 15 introduces the supervisory control problem, that deals with solving the unknown component problem for the controller's topology, and must also take into account partial controllability (and partial observability). We show how to model the problem in the framework of solving language equations for full controllability and partial controllability; for those cases we characterize controllers that are solutions. We also introduce the notion of weak controllers that model an extended class of controllers with respect to a looser notion of partial controllability that has a natural interpretation.

Chapter 16 addresses the problem of testing a component embedded within a modular system [131], that is known as the problem of testing in context or embedded testing. Solving this problem, the modular system is conveniently represented as two communicating machines, the embedded component machine, and the context machine that models the remaining part of the system which is assumed to be correctly implemented.

Chapter 17 discusses the application of language solving to games. First we summarize some of the existing results in game theory. Then we show how the theory of this book and the BALM program can be applied to games. Typically the rules of the game are described in the fixed part of the description which also keeps track of the game state. In addition, it is usually convenient to have an output which indicates if the game has been won and by whom. This reduces the specification to a simple automaton which monitors this output. The unknown component represents one players winning strategy, which is to be solved for if it exists. The other player is represented by unconstrained inputs. To restrict this player to legitimate moves, we map all illegal moves as a win for the first player. However, a winning strategy (or at least a non-losing strategy) for the first player is one where he must be able to win (or tie) no matter what (legal) move the second player makes. The game of tic-tac-toe is used to illustrate how a game can be modeled BLIF-MV and solved using BALM. The techniques described in this chapter generally apply to finite games where after a finite number of moves, the game is over. However, if a strictly winning strategy is desired, it might be necessary to model the game with ω-automata (as discussed in Chap. 18).

Chapter 18 illustrates how, in certain situations, the theory and implementation developed for finite-word automata can be applied to infinite-word automata. Although the theory of language solving can be extended formally (as discussed in Chap. 4 in Part I), the complexity of solving becomes doubly exponential in general. In this chapter, a restricted form of ω-language solving is discussed, which almost fits into the finite-word implementation, as available in BALM, except for a small modification of the derived result at the end.

Index

T. Villa et al., *The Unknown Component Problem: Theory and Applications*,
DOI 10.1007/978-0-387-68759-9, © Springer Science+Business Media, LLC 2012

R

S

W